Student Solutions Manual

to accompany

Analytical Chemistry

Seventh Edition

Gary D. Christian, Purnendu K. Dasgupta, and Kevin A. Schug

by

Gary D. Christian
University of Washington

WILEY

Cover Photo Credit: © yxowert/Shutterstock

ISBN-13   978-1-118-75209-8

Printed in the United States of America

V10003376_081518

Printed and bound by Quad/Graphics.

PREFACE

This Solutions Manual contains the answers to all Questions and solutions to all Problems in the textbook. Solutions to problems should be attempted before referring to the Solutions Manual. There is frequently more than one way to set up a problem, and your setup may differ from the one illustrated here, which is fine so long as you get the right answer. The detailed solutions given here are intended as a guide to help you get started when necessary. The final answers to the problems are given in Appendix F of the textbook.

Spreadsheet problems

The solutions to all spreadsheet problems are given on the textbook website.

TABLE OF CONTENTS

CHAPTER 1   ANALYTICAL OBJECTIVES, OR: WHAT ANALYTICAL CHEMISTS DO

1. The chemical characterization of matter.

2. Qualitative analysis deals with the identification of the presence of a particular substance or substances in a sample. Quantitative analysis deals with determining how much is present.

3. Define the problem, obtain a representative sample, dry the sample if required, measure its weight or volume, dissolve the sample and prepare the solution for the measurement step, measure the analyte, calculate the amount or concentration of analyte in the sample, and compute the precision of the analysis.

4. A sample represents the material to be analyzed. The analyte is the substance to be measured or determined. Hence, we determine the analyte by analyzing the sample.

5. A blank consists of all chemicals used in an analysis, run though the analytical procedure, to determine impurities that might be added to the analytical result, and which must be subtracted.

6. Gravimetry, volumetric analysis, instrumental analysis, kinetic methods of analysis, and combinations of these.

7. Precipitation (gravimetry), chromatography, solvent extraction, volatilization (distillation).

8   The measurement of a physical property of the sample.

9. A calibration curve represents an instrument (detector) response as a function of concentration. It may be a linear or a nonlinear response. An unknown analyte concentration in a sample solution is determined by comparison of the response with the calibration curve.

10. A specific reaction occurs only with the substance (analyte) of interest. A selective reaction occurs preferentially with the substance of interest, but not exclusively.

11. (a) Precipitate chloride with silver nitrate and weigh the purified precipitate.
       Measure sodium by atomic spectroscopy or an ion-selective electrode to distinguish form KCl impurities.
    (b) Titrate with standard sodium hydroxide solution.
    (c) Measure potentiometrically with a pH meter/electrode.

CHAPTER 2   BASIC TOOLS AND OPERATIONS OF ANALYTICAL CHEMISTRY

1. Volumetric (t.c.), pipets (t.d) (some micropipets t.c.), burets (t.d.).

2. The electronic balance is based on the principle of electromagnetic force compensation. The sample pan is placed on a movable hanger. The position of the hanger is monitored by an electrical position scanner, and a compensation current, proportional to the mass placed on the pan, brings it back to the zero postion. Older mechanical balances are first class levers in which an unknown mass is balanced against a known mass. If each arm of the lever is equal in length, then the two masses at balance are equal. The single-pan mechanical balance is a first class lever, but has unequal lever lengths, and is operated by removing weights on the sample end equal in value to the sample weight (see the text website).

3. Because lighter materials are used, the ratio of beam mass to length is decreased, the center of gravity is adjusted for greater sensitivity, the pan mass is decreased, and so forth.

4. The "TD" means "to deliver", and "TC" means "to contain" the specified volume.

5. The sample plus container is weighed, the sample is removed, and the loss in weight is the weight of the sample. This technique is useful for weighing hygroscopic samples that must be kept stoppered and for weighing several successive aliqouts of the same sample.

6. Don't handle objects with the fingers, weigh objects at room temperature with the balance door closed, never place chemicals directly on the pan, and frequently check the zero setting of the balance.

7. Concentrated hydrochloric acid is diluted, preferably with boiled distilled water. It is standardized by titrating against primary standard sodium carbonate or tris(hydroxymethyl)aminomethane. A saturated solution of sodium hydroxide is prepared and the insoluble sodium carbonate is allowed to settle out and then the supernatant is decanted. Or the saturated solution is filtered. The solution is diluted with boiled distilled water and standardized against primary standard potassium acid phthalate.

8. Dry ashing involves burning away the organic matter at an elevated temperature ($400\text{-}700^0$ C) with atmospheric oxygen as the oxidant. In wet digestion, the organic matter is oxidized to CO_2, H_2O, and other products by a hot oxidizing acid. Dry ashing is relatively free from contamination, but it has the danger of loss by volatilization or retention. Wet digestion is relatively free from retention and volatility losses, but it has the danger of contamination from impurities in the reagents.

9. Acid dissolution and acid or alkaline fusion followed by acid, neutral, or alkaline dissolution.

10. Protein free filtrate. It is prepared by mixing a biological fluid with a protein precipitating agent, such as trichloroacetic acid, tungstic acid, barium sulfate, etc., followed by filtering or centrifuging the precipitated proteins.

11. Care must be taken to prevent the digestion mixture from going to near dryness. Perchloric acid must not be added directly to organic or biological material, but only; after an excess of nitric acid is added. The fumes from the digestion should be collected or else a specially designed hood used.

12. The gross sample is the entire collected sample that is representative of the whole. This is reduced to a size suitable for handling, called the sample. An aliquot of the sample, called the analytical sample, is weighed and analyzed. Several aliquots of the sample may be analyzed. A grab sample is a single random sample that is assumed to be representative of the whole, an assumption that is valid only for homogeneous samples.

13. The electric field of the microwave energy causes molecules with dipole moments to rotate to try to align with the electric field, and ions migrate in the electric field. These movements result in heat.

14. Weight in air or water contained:

$$\begin{array}{r} 52.127 \text{ g} \\ -27.278 \text{ g} \\ \hline 24.849 \text{ g} \end{array}$$

$W_{vac} = 24.849 + 24.849\ (0.0012/1.0 - 0.0012/7.8)$

$\qquad = 24.849 + 0.026 = 24.875$ g

$V_{22}^{0} = 24.875$ g$/0.99777$ g/mL $= 24.931$ mL

$V_{20}^{0} = 24.931$ mL $(0.99777/0.99821) = 24.920$ mL

15. From Table 2.4:

$V_{25}^{0} = 24.971 \times 1.0040 = 25.071$ Ml

$V_{20}^{0} = 25.071$ mL $\times (0.9970/0.9982) = 24.041$ mL

16. From Table 2.4, the volumes expand by the ratio of $1.0054/1.0028 = 1.0026$. So (volumes are all in mL):

Nom. vol.	Vol. 20^{0}	Vol. 30^{0}	Change	Correction 30^{0}
10	10.02	10.05	+0.03	+0.05
20	20.03	20.08	+0.05	+0.08
30	30.00	30.08	+0.08	+0.08
40	39.96	40.06	+0.10	+0.06
50	49.98	50.11	+0.13	+0.11

17. From Table 2.4:

0.05129 x 0.9980/0.9959 = 0.05140 M

The volume expansion is 0.21%, causing the concentration to decrease this amount.

18. (b)

19. $W_{vac} = W_{air} + W_{air} (D_{air}/D_{object} - D_{air}/D_{weights})$
    $= 15.914 \text{ g} + 15.914 \text{ g} (0.0012 \text{ g/mL}/2.16 \text{ g/mL} - 0.0012 \text{ g/mL}/7.8 \text{g/mL})$
    $= 15.920 \text{ g}$

CHAPTER 3   STATISTICS AND DATA HANDLING IN ANALYTICAL CHEMISTRY

1. Accuracy is the agreement between the measured value and the accepted true value. Precision is the agreement between replicate measurements of the same quantity.

2. A determinate error is one that is non-random but is usually unidirectional and can be ascribed to a definite cause. An indeterminate error is a random error occurring by chance.

3. (a) determinate error, methodological
   (b) determinate error, methodological
   (c) indeterminate
   (d) determinate, instrumental

4. (a) 5 (b) 4 (c) 3

5. (a) 4 (b) 4 (c) 5 (d) 3

6. Li (6.9417) + N (14.0067) + 3O (46.9982 = 68.9466)

7. Pd (106.4) + 2Cl (70.9) = 177.3

8. $162._2$

9. $0.013_9 - 0.0067 + 0.00098 = 0.008_2$

10. To the nearest 0.01 g for three significant figures

11. (a) mean = 100 meq/L
    (b) absolute error = –2 meq/L
    (c) relative error = [–2/102] x 100% = –2%

12. (a) mean = 128.0 g
    (b) median = 128.1 g
    (c) range = 129.0 g – 127.1 g, or 1.9 g

13. (a) absolute error = 22.62 g – 22.57 g = 0.05 g
    relative error = (0.05 g/22.57 g) x 100% =0.2_2% =$2._2$ ppt

    (b) absolute error = 45.02 mL – 45.31 mL = –0.29 mL
    relative error = (–0.29 mL/45.31 mL) x 100% = –0.64% = –6.4 ppt

    (c) absolute error = 2.68% – 2.71% = –0.03%
    relative error = [(–0.03%)/(2.7%)] x 100% = $-1._1$% = -1_1 ppt

(d) absolute error = 85.6 cm − 85.0 cm = 0.6 cm

relative error = [(0.6 cm)/(85.0 cm)] x 100% = 0.7% = 7 ppt

14. (a) mean = 33.33% (see web spreadsheet calculations)

x_i	$x_i - \bar{x}$	$(x_i - \bar{x})^2$
33.27	0.06	0.0036
33.37	0.04	0.0016
33.34	0.01	0.00001
		Σ0.0053

$$s = \sqrt{(0.0053)/(3-1)} = 0.052\% \text{ (absolute)}$$

coeff. of varn = % rel. s = (0.052/33.33) x 100% = 0.16% (relative)

(b) mean = 0.024%

x_i	$x_i - \bar{x}$	$(x_i - \bar{x})^2$
0.022	0.002	4×10^{-6}
0.025	0.001	1×10^{-6}
0.026	0.002	4×10^{-6}
		$\Sigma 9 \times 10^{-6}$

$$s = \sqrt{(9x10^{-6})/(3-1)} = 0.0021\% \text{ (absolute)}$$

coeff. of varn = (0.0021/0.024) x 100% = 8.8% (relative)

15. mean = 102.6 (See web spreadsheet calculations)

x_i	$x_i - \bar{x}$	$(x_i - \bar{x})^2$
102.2	0.4	0.16
102.8	0.2	0.04
103.1	0.5	0.25
102.3	0.3	0.09
		Σ0.54

(a) $s = \sqrt{(0.54)/(4-1)} = 0.42$ ppm

(b) rel. s = (0.42/102.6) x 100% = 0.41%

(c) $s_{(mean)} = 0.42/\sqrt{4} = 0.21$ ppm

(d) rel. $s_{(mean)}$ = (0.21/102.6) x 100% = 0.20%

16. mean = 95.65% (See web spreadsheet calculations)

x_i	$x_i - \bar{x}$	$(x_i - \bar{x})^2$
95.67	0.02	0.0004
95.61	0.04	0.0016
95.71	0.06	0.0036
95.60	0.05	0.0025
		$\Sigma 0.0081$

(a) $s = \sqrt{(0.0081)/(4-1)} = 0.052\%$ (absolute)

(b) $s_{(mean)} = (0.052)/\sqrt{4} = 0.026\%$ (absolute)

(c) rel. $s_{(mean)} = [0.026/95.65] \times 100\% = 0.027\%$ (relative)

17. **PFP Solution**: We need to express the desired interval in multiples (z) of the standard deviation.

$$z = \frac{x - \mu}{s} = \frac{(50000 - 64700)}{6400} = -2.3$$

From the Standard Normal (Z) Table in www.statsoft.com/textbook/distribution-tables, the fraction of the area under the Gaussian curve between the average value (the maximum of the curve) and z = –2.3 is 0.4893. This is in the intersecting cell for 2.30 and 0.00. (Compare this with –2σ in Figure 3.2. In the Z table for –2.0, the area is 0.4772, which for –2σ to +2σ is 0.9944 or 95.44%).

The entire area from –∞ to z = –2.3 will be 0.5000–0.4893 = 0.0107

(a) Therefore the fraction of brakes to be 80% worn through in less than 50,000 miles is 1.07%.

(b) (1,000,000 × 1.07%) = 10,700 extra brakes should be kept available for every 1 million product sold.

18. (a) $s_a = [(\pm 2)^2 + (\pm 8)^2 + (\pm 4)^2]^{1/2} = (\pm 84)^{1/2} = \pm 9.2$
$128 + 1025 - 636 = 517 \pm 9$

(b) $s_a = [(\pm 0.06)^2 + (\pm 0.03)^2]^{1/2} = (\pm 0.0045)^{1/2} = \pm 0.067$
$16.25 - 9.43 = 6.82 \pm 0.07$

(c) $s_a = [(\pm 0.40)^2 + (\pm 1)^2]^{1/2} = (\pm 1.16)^{1/2} = \pm 1.1$
$46.1 + 935 = 981 \pm 1$

(See text website for spreadsheet calculations of s_a)

19. (a) $(2.78 \pm 0.04)(0.00506 \pm 0.00006) = 0.01407 \pm ?$
$(s_b)_{rel} = (\pm 0.04)/(2.78) = \pm 0.014$
$(s_c)_{rel} = (\pm 0.00006)/(0.00506) = \pm 0.012$
$(s_a)_{rel} = [(\pm 0.014)^2 + (\pm 0.012)^2]^{1/2} = (\pm 0.00034)^{1/2} = \pm 0.018$
$s_a = (0.01407)(\pm 0.018) = \pm 0.0003$ (See web for spreadsheet calculation of $(s_a)_{rel}$)

(b) $(36.2 \pm 0.4)/(27.1 \pm 0.6) = 1.336 \pm ?$
$(s_b)_{rel} = (\pm 0.4)/(36.2) = \pm 0.011$
$(s_c)_{rel} = (\pm 0.6)/(27.1) = \pm 0.022$
$(s_a)_{rel} = [(\pm 0.011)^2 + (\pm 0.022)^2]^{1/2} = (\pm 0.00060)^{1/2} = \pm 0.024$
$s_a = (1.336)(\pm 0.024) = \pm 0.032$

Therefore, the answer is 1.34 ± 0.03

(c) $(50.23 \pm 0.07)(27.86 \pm 0.05)/(0.1167 \pm 0.0003) = 11,991 \pm ?$
$(s_b)_{rel} = (\pm 0.07)/(50.23) = \pm 0.0014$
$(s_c)_{rel} = (\pm 0.05)/(27.86) = \pm 0.0018$
$(s_d)_{rel} = (\pm 0.0003)/0.1167) = \pm 0.0026$
$(s_a)_{rel} = [(\pm 0.0014)^2 + (\pm 0.0018)^2 + (\pm 0.0026)^2]^{1/2} = \pm 0.0035$
$s_a = (11,991)(\pm 0.0035) = \pm 42$

Therefore, the answer is $11,990 \pm 40$ or $[(1.199 \pm 0.004) \times 10^4]$

(See text website for spreadsheet calculations)

20. $[(25.0 \times 0.0215 - 1.02 \times 0.112)(17.0)]/5.86$
$= (0.538 - 0.114)(17.0)/5.87 = (0.424)(17.0)/(5.87) = 1.228 \pm ?$

For 25.0×0.0215:
$(s_b)_{rel} = (\pm 0.1)/(25.0) = \pm 0.0040$
$(s_c)_{rel} = (\pm 0.0003)/(0.0215) = \pm 0.014$
$(s_a)_{rel} = [(\pm 0.0040)^2 + (\pm 0.014)_2]^{1/2} = \pm 0.015$
$s_a = (0.538)(\pm 0.015) = 0.0081$; Therefore, 0.538 ± 0.008

For 1.02×0.112:
$(s_b)_{rel} = (\pm 0.01)/(1.02) = \pm 0.0098$
$(s_c)_{rel} = (\pm 0.001)/(0.112) = \pm 0.0089$
$(s_a)_{rel} = [(\pm 0.0098)^2 + (\pm 0.0089)^2]^{1/2} = \pm 0.013$
$s_a = (0.114)(\pm 0.013) = \pm 0.0015$; Therefore, 0.114 ± 0.002

For $(0.538 \pm 0.008) - (0.114 \pm 0.002)$:
$s_a = [(\pm 0.0081)^2 + (\pm 0.0015)^2]^{1/2} = \pm 0.0082$; Therefore, 0.423 ± 0.008

For $[(0.423 \pm 0.008)(17.0 \pm 0.2)]/(5.87 \pm 0.01)$:

$(s_b)_{rel} = (\pm 0.0082)/(0.423) = \pm 0.019$

$(s_c)_{rel} = (\pm 0.2)/(17.0) = \pm 0.012$

$(s_d)_{rel} = (\pm 0.01)/(5.87) = \pm 0.0017$

$(s_a)_{rel} = [(\pm 0.019)^2 + (\pm 0.012)^2 + (0.0017)^2]^{1/2} = \pm 0.023$

$s_a = (1.225)(\pm 0.023) = \pm 0.028$

Answer $= 1.22 \pm 0.03$

21. PFP

(a) We propagate uncertainty through a summation by taking the square root of the sum of the squares of individual absolute uncertainties.

$$S_{total} = \sqrt{\left(s_{GHG}^2 + s_{ozone}^2 + s_{surfacealbedo}^2 + s_{aerosoldirect}^2 + s_{aerosolindirect}^2\right)}$$

$$S_{total} = \sqrt{\left(0.26^2 + 0.22^2 + 0.2^2 + 0.36^2 + 0.7^2\right)}$$

$$S_{total} = 0.88 \ \text{W/m}^2$$

The terms for the aerosols seem to contribute most to the uncertainty. In fact if only the aerosol terms were considered the uncertainty would still be 0.79 W/m^2. Often, propagated uncertainty is dominated by the least precisely known values.

(b) $\Delta T_{surface} = \lambda \times \Delta F_{net} = 0.7 \times 1.61 = 1.13K$

$$\frac{s_a}{a} = \sqrt{\left(\frac{s_x}{x}\right)^2 + \left(\frac{s_y}{y}\right)^2}$$

$$\frac{s_a}{1.13} = \sqrt{\left(\frac{0.4}{0.7}\right)^2 + \left(\frac{0.88}{1.61}\right)^2}$$

$$\frac{s_a}{1.13} = 0.79$$

$$s_a = 0.89K$$

This yields an expected mean temperature change of 1.13 ± 0.89 K.

(b) Yes, the instrumental record is well within the range specified by the calculation. Furthermore, the instrumental record of temperatures only dates to the 1850's while the radiative forcing estimates reference the year 1750.

22. mean = 0.05027 M (See website for spreadsheet calculation of s)

x_i	$x_i - \bar{x}$	$(x_i - \bar{x})^2$
0.5026	0.0001	1×10^{-8}
0.5029	0.0002	4×10^{-8}
0.5023	0.0004	16×10^{-8}
0.5031	0.0004	16×10^{-8}
0.5025	0.0002	4×10^{-8}
0.5032	0.0005	25×10^{-8}
0.5027	0.0000	0
0.5026	0.0001	1×10^{-8}
		$\Sigma 67 \times 10^{-8}$

$$s = \sqrt{(67 \times 10^{-8})/(8-1)} = 0.00031 \ M$$

From Equation 2.9 and Table 2.1 (t = 2.365 for υ = 7 at 95% C.L.):

Conf. limit = $0.5027 \pm (2.365 \times 0.00031)/\sqrt{8}$
         = $0.5027 \pm 0.00026 \ M$, or $0.5024 - 0.5030 \ M$

23. PFP

	Brand A	Brand B	Brand C	Brand D
average	5.58	5.52	5.70	6.91
stdev	0.19	0.08	0.29	0.07
t at 95% CI				
(from t-table)	2.262	2.262	2.262	2.262
calculated CI	0.134034	0.056424	0.2078635	0.05278
Avg+/- CI	5.6 ± 0.1	5.52 ± 0.06	5.7 ± 0.2	6.91 ± 0.05

Brand B gives the most accurate reading, and with good precision.

24. mean = 139.6 meq/L (See website for spreadsheet calculation of s)

x_i	$x_i - \bar{x}$	$(x_i - \bar{x})^2$
139.2	0.4	0.16
139.8	0.2	0.04
140.1	0.5	0.25
139.4	0.2	0.04
		$\Sigma 0.49$

$$s = \sqrt{(0.49)/(4-1)} = 0.40 \ \text{meq/L}$$

(a) For 3 degrees of freedom at the 90% confidence limit, t = 2.353

Conf. limit = $139.6 \pm (2.353 \times 0.40)/\sqrt{4}$

$= 139.6 \pm 0.47$ meq/L, or $139.1 - 140.1$ meq/L

(b) t = 3.182 at 95% confidence level

Conf. limit = $139.6 \pm (3.182 \times 0.40)/\sqrt{4}$

$= 139.6 \pm 0.64$ meq/L, or $139.0 - 140.2$ meq/L

(c) t = 5841 at 99% confidence level

Conf. limit = $139.6 \pm (5.841 \times 0.40)/\sqrt{4}$

$= 139.6 \pm 1.17$ meq/L, or $138.4 - 140.8$ meq/L

Note that in order to be more confident of the range of the true value, the range must increase. Conversely, as the range is narrowed, we are less confident that it defines the true value.

25. t for 2 degrees of freedom at 90% confidence level = 2.920

Conf. limit = $\pm(2.920 \times 2.3)/\sqrt{3} = \pm 3.9$ ppm

26. PFP

First, calculate the average of normal cell count,

$$\frac{(5.2 + 4.8 + 5.4 + 5.3 + 5.1 + 4.9 + 5.5)}{7} = 4.7_4$$

Calculate the standard deviation, s = 0.83_0

For 95% confidence interval,

$$\mu = \bar{x} \pm \frac{ts}{\sqrt{N}} = 4.7_4 \pm \frac{2.447 \times 0.083}{\sqrt{7}} = 4.7_4 \pm \frac{2.447 \times 0.083}{\sqrt{7}} = 4.7_4 \pm 0.76_8$$

A cell count of 4.5 million falls in the range of 95% confidence interval, and therefore it is not too low.

27. t for 2 degrees of freedom at 95% confidence level = 4.303

Conf. limit = $\pm(4.303 \times 0.5)/\sqrt{3} = \pm 1._2$ meq/L

28. PFP

The average and standard deviation of the analysis are 95.4 and 3.0_3, respectively. Therefore, 95% confidence interval = 90.6 to 100.2. This analysis is within the 95% confidence interval and therefore not significantly different from the known value.

29. Applying the Q-test to the standardization data shows that 0.1050 should probably be rejected. Then the valid data are 0.1071, 0.1067, and 0.1066. The standard deviation for these triplicate results calculates to be 0.00026 M (mean 0.1068M). $t = 2.920$

   Conf. limit $= 0.1068 \pm (2.920 \times 0.00026)/\sqrt{3}$
   $= 0.1068 \pm 0.0004\ M$, or $0.1064 - 0.1072\ M$

30. PFP

   (a) Q_{90} for $n = 5$ is 0.64
       The questionable data point is 3.16.
       Q = the gap between the two most widely-spaced consecutive data points/the range between the highest and lowest data points $= (3.16 - 3.07)/(3.16 - 2.98) = 0.5 < 0.64$
       Because our calculated value is lower than the statistical value from the table, we cannot discard 3.16.

   (b) 2.776 is the t value at the 95% confidence level for 4 degrees of freedom.
       The mean value for the data points $= 3.04_4$
       The standard deviation for this data set $= 0.07_4$
       The confidence interval for this data set $= 3.04_4 \pm (2.776 \times 0.07_4)/(5^{1/2}) = 3.04_4 \pm 0.09_2$
       Because the true value of 3.03 falls within the confidence range (2.95_2 to 3.13_6), we can be 95% confident that our results agree with the known value.

31. $\bar{X}_C = 15._8$; $\bar{X}_D = 23._3$

$(X_i - \bar{X}_C)^2$	$(X_i - \bar{X}_D)^2$
1	4
49	9
16	144
4	81
49	64
144	289
25	49
36	169
$\Sigma 324$	1
	25
	$\Sigma 835$

F-test: $F = (s_D{}^2)/(s_C{}^2) = (835/9)/(324/7) = 2.0_0$ (See website for spreadsheet calculation) $F_{Table} = 3.68$. Therefore, the precision of the two groups is comparable and the t-test can be applied. Use the paired t-test.

   $S_\mu = \sqrt{(324 + 835)/(8 + 10 - 2)} = 8.51$

   $t_{calc} = (23._3 - 15._8)/8.51 \sqrt{(10 x 8)/(10 + 8)} = 1.8_6$

This is smaller than the tabulated t value for 16 degrees of freedom at the 95% confidence level, but not at the 90% confidence level. It appears there is a fair probability the differences between the two populations is real. More studies are indicated.

32. PFP

subject	before	after	d
1	312	300	−12
2	242	201	−41
3	340	232	−108
4	388	312	−76
5	296	220	−76
6	254	256	2
7	391	328	−63
8	402	330	−72
9	290	231	−59

ave	−56.1111
std dev	34.17398
t=	4.925774
t table 99%	3.355

Yes, the differences are significant.

33. $\overline{X}_E = 13.0_0$; $\overline{X}_G = 13.1_8$

$(x_i - \overline{X}_E)^2$	$(x_i - \overline{X}_G)^2$
0.01	0.09
0.09	0.01
0.16	0.04
0.09	0.09
0.09	Σ0.23
Σ0.44	

First perform an F-test.

$F = (s_E{}^2)/s_G{}^2) = (0.44/4)/(0.23/3) = 1.4_3$ (See website for spreadsheet calculation)
$F_{table} = 9.12$. Hence, there is a high probability the variances of the two methods represent the same population variance (note the standard deviations are nearly identical).

Apply Equation 3.14:

$$S_p = \sqrt{(0.44 + 0.23)/(9 - 2)} = 0.31$$

$t_{calc} = [(13.0_0 - 13.1_8)/0.31]\sqrt{(5x4)/(5+4)} = 0.8_7$

$t_{table} = 2.365$ for 7 degrees of freedom ($(N_1 + N_2 - 2)$, so there is a high probability othe two methods give the same result.

34.

$(x_i - \overline{X_A})^2$	$(x_i - \overline{X_B})^2$	$(x_i - \overline{X_C})^2$ (See text website for spreadsheet
0.00036	0.00026	0.00102 calculaltion)
0.00122	0.00012	0.00029
0.00122	0.00002	0.00040
0.00040	Σ0.00040	Σ0.00173
Σ0.00320		

$$s_p = \sqrt{(0.0032 + 0.00040 + 0.00173)/(11-3)} = 0.0258 \text{ absorbance units}$$

35. Colorimety: $s_1^2 = [\Sigma(D_i - \overline{D})^2]/(N-1) = 0.933$

AAS: $s_2^2 = (1.67)/(6-1) = 0.334$

$F = (s_1^2)/(s_2^2) = (0.933)/(0.334) = 2.79$ (See website for spreadsheet calculation)

F_{table} for $v_1 = 7$ and $v_2 = 5$ is 4.88. Since $F_{calc} < F_{table}$, there is no significant difference in the two variances.

36. PFP

To evaluate method accuracy, students should compare the certified value for SRM 872 to the measured values for the SRM. Only one replicate analysis for each method was performed, so a t test can not be done. The Hot Plate Method result was slightly higher than the certified value range. In comparing the other Hot Plate Method and Digestion Block Method data for each coin half (1a, 1b, 2a, etc.), there does not appear to be a bias in the results. The Digestion Block method result was within the certified range for SRM 872. Replicate analyses using each digestion method would aid in our evaluation of accuracy.

To evaluate precision, data for each coin can be compared. There appears to be poor precision for the replicate data within each method (Coin 1 Hot Plate: 14.44% vs. 8.41%; Coin 2 Hot Plate: 23.77% vs 27.23%, etc). This poor precision is likely due to a lack of homogeneity of lead in the bronze. Averaging the replicate data for each coin would provide a better overall indication of each coin's lead content. Additional replicate analyses of each coin also would be beneficial to further investigate the homogeneity of lead in the bronze. In comparing the lead contents among the four coins, the results vary from 6.34% to 27.23%. This demonstrates that the lead contents of individual ancient bronze coins can vary significantly, as stated in the problem description.

You should perform a paired t-test to evaluate whether these is a statistical difference between the two methods. The results are shown below. No statistical difference between the methods is indicated at a 95% confidence level.

Coin	Hot Plate Method (%)	Digestion Block Method (%)	Difference (%)
1a	14.44	15.37	−0.93
1b	8.41	7.24	1.17
2a	23.77	24.14	−0.37
2b	27.23	24.87	2.36
3a	6.34	6.77	−0.43
3b	8.04	7.34	0.7
4a	16.16	17.2	−1.04
4b	19.07	18.26	0.81
NIST SRM 872	4.21	4.12	0.09

Average	0.26222222
Standard Deviation	1.10352591
n	9
t_{table} (df=8, P=0.05)	2.31
t_{calc}	0.71286651

$t_{calc} < t_{table}$ No difference between methods at 95% confidence level

An Excel file for the solution is in the web supplement.

37. The mean is $0.1017\ M$ and the standard deviation is $0.0001_7\ M$

$t_{calc} = (\overline{x} - \mu)/(\sqrt{N}/s) = (0.1017 - 0.1012)(\sqrt{4}/0.0001_7) = 5._9$ (See website for spreadsheet calculation)

This exceeds the tabulated t value, even at the 99% confidence level, so there is a 99% probability that the difference is real and not due to chance.

38. The mean is 99.89% with a standard deviation of 0.033%.

$t_{calc} = (\overline{x} - \mu)/(\sqrt{N}/s) = (99.89 - 99.95)(\sqrt{4}/0.033) = 3._6$ (See website for spreadsheet calclation)

This just exceeds the tabulated t value at the 95% confidence level. Hence, there is a 95% probability that the analyzed data are significantly different from the supplier's stated value. Note that the difference of 0.060% is about twice the standard deviation, and we would expect this to occur by chance only 1 out of 20 (5%). Whether the shipment is accepted depends on the acceptable differences.

39. PFP

(a)
   i. calculate the mean
   ii. calculate the standard deviation
   iii. calculate t for your set of measurements in comparison to known value
   iv. compare to t values from the t-table
      a. if $t_{calculated}$ is greater than t_{table} we can be 95% confident that the two values are significantly different
      b. if $t_{calculated}$ is less than t_{table} we can be 95% confident that the two values are statistically the same

Mean: $\overline{x} = \dfrac{181.83 + 182.12 + 182.32 + 182.20}{4} = 182.12 \; mg/dL$

Standard deviation:

$$s = \sqrt{\dfrac{(181.83 - \overline{x})^2 + (182.12 - \overline{x})^2 + (182.32 - \overline{x})^2 + (182.20 - \overline{x})^2}{4 - 1}} = 0.208$$

So your measurement: $182.1_2 \pm 0.2_1$ mg/dL

Calculate t:

$$t_{calculated} = \dfrac{\left| \overline{x} - known \; value \right|}{s} \sqrt{n} = \dfrac{\left| 182.12 - 182.15 \right|}{0.208} \sqrt{4} = 0.2857$$

Compare to t_{table}:
t_{table} for 3 d.f. at 95% confidence level: 3.182

So $t_{calc} < t_{table}$ so our measurement result is the same as the reported NIST standard.

(b) To compare two sets that BOTH have errors, you have to calculate the means and standard deviations of each, and "pool" the standard deviations:

$\overline{x} = 146.26$ mg/dL $\overline{x} = 146.29$ mg/dL
$s = 0.08_3$ $s = 0.1_0$

$$S_{pooled} = \sqrt{\dfrac{\underset{set \; 1}{\sum}(x_i - \overline{x}_i)^2 + \underset{set \; 2}{\sum}(x_i - \overline{x}_i)^2}{n_1 + n_2 - 2}} = 0.09375$$

NOW we can calculate t:

$$t_{calculated} = \dfrac{\left| \overline{x}_1 - \overline{x}_2 \right|}{s_{pooled}} \sqrt{\dfrac{n_1 n_2}{n_1 + n_2}} = 0.1624$$

Then, compare to value in t-table:

t_{table} (at 95% confidence level, N=10) = 2.228

If t_{calc} is greater than t_{table} we can be 95% confident that the two measurement

Methods produce DIFFERENT values.

For our data: $t_{calc} < t_{table}$ so our measurement method gives the same results as the accepted lab method.

(c) We first compare the individual differences:

#1	174.60 – 174.93 =	–0.33
#2	142.32 – 142.81 =	–0.49
#3	210.67 – 209.06 =	1.61
#4	188.32 – 187.92 =	0.40
#5	112.41 – 112.37 =	0.04

Average difference, $\bar{d} = 0.246$ (keep signs when calculating average)

Now perform a t-test on the individual differences:

$$t_{calculated} = \frac{\bar{d}}{s_d}\sqrt{n} \qquad\qquad n = \text{number of pairs of data}$$

For us: n = 5

$$s_d = \sqrt{\frac{\sum(d_i - \bar{d})^2}{n-1}}$$

$$sd = \sqrt{\frac{(-0.33-0.246)^2 +(-0.49-0.246)^2 +(1.61-0.246)^2 +(0.40-0.246)^2 +(0.04-0.246)^2}{5-1}}$$

$s_d = 0.8367$

Now, calculate t:

$$t_{calculated} = \frac{\bar{d}}{s_d}\sqrt{n} = \frac{0.246}{0.8367}\sqrt{5} = 0.657$$

If t_{calc} is less than t_{table} then there is a 95% probability that the methods produce the same values.

$t_{table} = 2.776$ at 95% confidence level and N=4

So, your accepted method produces the same value as the accepted method within the experimental error of the methods! Let's patent your method!

40. Arrange in decreasing order:
    0.1071, 0.1067, 0.1066, 0.1050

    The suspect result is 0.1050
    $Q = (0.0016)/0.0021) = 0.76$

    Tabulated $Q = 0.829$.   Hence, it is 95% certain that the suspected value is not due to accidental error.

41. For the Zn determination:
    33.37, 33.34, 33.27

    The suspect result is 33.27
    $Q = (0.07)/(0.10) = 0.70$
    $Q_{table} = 0.970$ Therefore, the number 33.27 is valid.

    For the Sn determination:
    0.026, 0.025, 0.022

    The suspect result is 0.022
    $Q = (0.002)/(0.004) = 0.75$
    $Q_{table} = 0.970$ Therefore, 0.022 is a valid number

42. Arranging in order:
    22.25, 22.23, 22.18, 22.17, 22.09
    $Q = (0.08)/(0.16) = 0.50$
    $Q_{table} = 0.710$. Therefore, 22.09 is a valid measurement.

43. PFP

    (a) First, sort the values in an increasing order. Q value at 90% confidence for 6 observations is 0.56. The calculated Q values 0.31 and 0.51 are less than 0.56 and therefore all data points should be kept.

41.99		
43.15		
43.23		
43.56		
43.81		
45.71		
3.72	Range	Q
1.16	Gap 1	0.311828
1.9	Gap 2	0.510753

    (b) (43.58 ± 1.22) %

(c) First, we perform F test to check whether two standard deviations are significantly different from each other.

$$F = \frac{s_1^2}{s_2^2} = \frac{(1.22)^2}{(0.44)^2} = 7.67 > F_{table} (= 6.26)$$

$$t_{calculated} = \frac{|\bar{x}_1 - \bar{x}_2|}{\sqrt{\dfrac{s_1^2}{n_1} + \dfrac{s_2^2}{n_2}}}, \text{ degree of freedom} = \left\{ \frac{\left(\dfrac{s_1^2}{n_1} + \dfrac{s_2^2}{n_2}\right)^2}{\dfrac{\left(\dfrac{s_1^2}{n_1}\right)^2}{n_1 + 1} + \dfrac{\left(\dfrac{s_2^2}{n_2}\right)^2}{n_2 + 1}} \right\} - 2$$

$t_{calculated} = 1.82$; $df = 7.10 \approx 7$

$t_{table} = 2.365$

$t_{calculated} < t_{table}$, the difference is not significantly different at 95% confidence, even at 99% confidence level. The suspect is highly likely to be the person who committed the crime.

44. The range is $103.1 - 102.2 = 0.9$ ppm
    From Equation 3.17 and Table 3.4 for 4 observations,
    $s_r = (0.9)(0.49) = 0.44$ ppm
    This compares with $s = 0.42$ ppm calculated in Problem 15

45. The range is $0.5032 - 0.5023 = 0.0009\ M$
    From Equation 3.18 and Table 3.4 for 8 observations, conf. limit $= 0.5027 \pm 0.0009\ (0.029) =$
    0.5027 ± 0.00026 or $0.5024 - 0.5030$.
    This is identical to the confidence limit calculated using the standard deviation.

46. The range is $140.1 - 139.2 = 0.9$ meq/L
    From Equation 3.18 and Table 3.4 for 4 observations, conf. limit$_{95\%} = 139.6 \pm 0.9\ (0.72) =$
    139.6 ± 0.65 or $139.0 - 140.2$, the same as using the standard deviation. conf. limit$_{99\%} =$
    $139.6 \pm 0.9\ (1.32) = 139.6 \pm 1.19$ or $138.4 - 140.8$, the same as using the standard deviation.

47.

$(x_i - \bar{X})$	$(x_i - \bar{X})^2$	$(x_i - \bar{Y})$	$(x_i - \bar{X})(x_i - \bar{Y})$
−0.300	0.0900	−16.7	5.01
−0.200	0.0400	−10.9	2.18
−0.100	0.0100	−4.5	0.45
0.100	0.0100	5.6	0.56
0.500	0.2500	26.6	13.3_0
	$\Sigma 0.4000$		$\Sigma 21.5_0$

   $m = (21.50)/(0.4000) = 53.7_5$ (See text website for spreadsheet calculation)
   This is identical to the value obtained with Equation 3.23. See Example 3.21.

48.

x_i	y_i	x_i^2	$x_i y_i$
1.00	0.205	1.00	0.205
2.00	0.410	4.00	0.820
3.00	0.615	9.00	1.84_5
4.00	0.820	16.00	3.28_0
$\Sigma 10.00$	$\Sigma 2.050$	$\Sigma 30.00$	$\Sigma 6.15_0$

$(\Sigma x_i)^2 = 100.0$

$\overline{X} = (\Sigma x_i)/n = 2.500$ $\overline{y} = (\Sigma y_i)/n = 0.5125$ $n = 4$

Using Equations 3.23 and 3.22:

$m = [6.15_0 - (10.00 \times 2.050)/4]/30.0_0 - 100.0/4) = 0.205$

$b = 0.5125 - (0.205)(2.500) = 0.000$

$y = 0.205x + 0.00$

Unknown:

$0.625 = 0.205x + 0.000$

$x = 3.05$ ppm P in urine

(See text website for spreadsheet for plot and calculation)

49. From Problem 48:

$Sy_i^2 = (0.205)^2 + (0.410)^2 + (0.615)^2 + (0.820)^2 = 1.260_7$

$(Sy_i)^2 = (2.050)^2 = 4.202$

$Sx_i^2 = 30.00;\ (Sx_i)^2 = 100.0;\ m^2 = (0.205)^2 = 0.0420$

From Equation 3.24,

$s_y = \sqrt{[(1.260_7 - 4.202/4) - 0.0420(30.00 - 100.04/4)]/(4-2)}$

$= \pm 0.01_0$ absorbance

From Equation 3.25,

$s_m = \sqrt{(0.01_0)/(30.00 - 100.00/4)} = \pm 0.004_5$ absorbance/ppm

$m = 0.205 \pm 0.004$

From Equation 3.26

$s_b = 0.01_0 \sqrt{30.00/[4(30.00) - 100.00]} = \pm 0.01_2$ absorbance

$b = 0.00_0 \pm 0.01_2$

The phosphorus concentration in the urine sample is given by

$x = [(0.62_5 \pm 0.01_0) - (0.00_0 \pm 0.00_1)]/(0.205 \pm 0.004) = 3.05 \pm ?$

$s_{num} = \sqrt{\pm(0.01_0)^2 + (0.00_1)^2} = \pm 0.01_0$

$(s_{div})_{rel} = \sqrt{\pm(0.01_0/0.62_5)^2 + (\pm 0.004/0.205)^2} = \pm 0.02_5$

$s_{div} = 3.05\ (\pm 0.02_5) = \pm 0.07_6$

$x = 3.05 \pm 0.08$ ppm P in urine

50. PFC
The equations will fail if the errors are not randomly distributed, if there are more outliers, for example. They will also fail if the functional form that the data should exhibit is not a straight line.

51.

% yeast extract (x_i)	Toxin, mg (y_i)
1.000	0.487
0.200	0.260
0.100	0.195
0.010	0.007
0.001	0.002
$\Sigma 1.311$	$\Sigma 0.951$

$\overline{x_i} = 1.311/5 = 0.262 \qquad \overline{y_i} = 0.951/5 = 0.190$

$\Sigma x_i^2 = 1.050 \qquad\qquad \Sigma y_i^2 = 0.343$
$n = 5 \qquad\qquad\qquad \Sigma x_i y_i = 0.559$

$r = (n\,\Sigma x_i y_i - \Sigma x_i\,\Sigma y_i)/\{[n\Sigma x_i^2 - (\Sigma x_i)^2][n\Sigma y_i^2 - (\Sigma y_i)^2]\}^{1/2}$
$= (2.795 - 1.247)/[(5.250 - 1.719)(1.715 - 0.904)]^{1/2} = 0.915$ (See website)

$r^2 = 0.84$ (see website for spreadsheet calculation)

Therefore, there is a good correlation between yeast extract concentration and the amount of toxin produced.

52.

Toxin, mg (x_i)	Dry weight (y_i)
0.487	116
0.260	53
0.195	37
0.007	8
0.008	1
$0.951 = \Sigma x_i$	$215 = \Sigma y_i$

$\overline{x_I} = 0.951/5 = 0.190 \qquad \overline{y_i} = 215/5 = 43$

$\Sigma x_i^2 = 0.343 \qquad\qquad \Sigma y_i^2 = 17699$
$\Sigma x_i y_i = 77.5$

$r = (387/5 - 204.5)/\sqrt{(1.715 - 0.904)(88495 - 46225)}$

$= (183.0)/\sqrt{(0.811)(42270)} = (183.0)/185) = 0.989; \, r^2 = 0.978$

(See website for spreadsheet calculation)

Therefore, there is a strong correlation between fungal dry weight and amount of toxin produced.

53. Enzyme   Colori-   (See website for spreadsheet calculations)
    method   metric
               method

(x_i)	(y_i)	D_i	$D_i - D$	$(D_i - D)^2$	$x_i y_i$	x_i^2	y_i^2
305	300	5	1	1	91,500	93,025	90,000
385	392	−7	−11	121	150,920	148,225	153,664
193	185	8	4	16	35,705	35,705	34,225
162	152	10	6	36	24,624	26,624	23,104
478	480	−2	−6	36	229,440	228,484	230,400
455	461	−6	−10	100	209,755	207,025	212,521
238	232	6	2	4	55,216	56,644	53,824
298	290	8	4	16	86,420	88,420	84,100
408	401	7	3	9	163,608	166,464	160,801
323	315	8	4	16	101,745	104,745	99,225
$\Sigma3,245$	$\Sigma3,208$	$\Sigma37$		$\Sigma355$	$\Sigma1,148,933$	$\Sigma156,493$	$\Sigma1,141,864$

$D = 3.7$

From Equation 3.16:

$$s_d = \sqrt{(355)/(10-1)} = 6.28$$

From Equation 3.15:

$$t = (3.7/6.28)\sqrt{10} = 1.8_6$$

From Table 3.1 at the 95% confidence level and $\upsilon = 9$, $t = 2.262$. Since $t_{calc} < t_{table}$, there is no significant difference between the methods at the 95% confidence level.

From Equation 3.28, we calculate:

$$r = [10(1,148,933) - (3,245)(3,208)]/\{[10(1,156,493) - (3,245)^2][10(1,141,864) - (3,208)^2]\}^{1/2} = 0.999; \, r^2 = 0.998$$

Hence, there is a high degree of correlation.

54. The average blank reading is 0.18, and the standard deviation is ±0.06.
    The net reading for the detection limit is 3 x 0.06 = 0.18
    The net reading for the standard is 1.25 − 0.18 = 1.07
    The detection limit is 1.0 ppm (0.18/1.07) = 0.17 ppm
    This would give a total (blank plus analyte) reading of 0.18 + 0.18 = 0.36

55. PFP

(a) Blank Average $S_{blank} = 0.0011_7$; Standard Deviation $\sigma = 0.0005_6$.
Minimum detectable signal, $y = S_{blank} + 3\sigma = 0.0028_5$

(b) Slope, $m = 0.7493$ 1/ppb

(c) Detection Limit $= (y - S_{blank})/m = 0.0023$ ppb
The calculation is correct, but a more accurate approach is to plot all the individual blank (0 concentrations) and non-blank corrected absorbances. Deleting the 2.5 ppb value gives a more linear plot in the DL region ($y = 0.8961x + 0.0006$, DL $= 0.0019$ppb). NOTE: See Chapt. 3 website, Documents, LOD calculation from calibration curve, from Peter Loock, for a more comprehensive way to calculate the LOD. See Sensors & Actuators B, 173 (2012) 157–163.

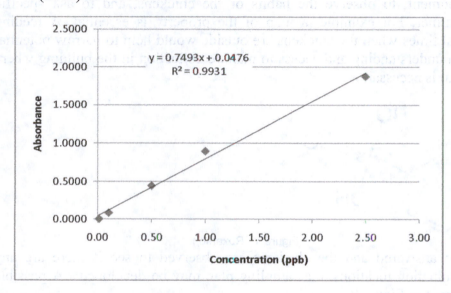

56. $WR^2 = K_s$
$K_s = (0.4\text{ g})(5)^2 = 10$ g
For R $= 0.25$,
$W(2.5)^2 = 10$ g
$W = 1.6$ g sample

57. PFP

(a) The probability of getting red M&M, $p = 200{,}000/(200{,}000 + 50{,}000) = 0.8$. The expected number of getting red M&M is $1000 \times p = 800$.

(b) Standard deviation in sampling operation $s = \sqrt{npq} = \sqrt{1000 \times 0.8 \times 0.2} = 12.6 \approx 13$. p and q are the probability of getting red and green M&M, respectively. $p = 0.8$ and $q = 0.2$.

58. $s_s = 0.15\%$ (wt/wt)
R $= 0.05$ and $s_x = (0.05)(3\%) = 0.15\%$ (wt/wt)
$x = 3\%$ (wt/wt)

From Equation 3.31,
$n = t^2 s_s^2 / R^2 x^2 = (1.96)^2 (0.15)^2 / (0.05)^2 (3)^2 = 3.84$

59. PFP

(a) First of all, focus specifically on the nature of the salmon. Salmon of various sizes, sex, type(Coho and Chinook), and age should be collected. These factors are important to examine as mercury accumulates in fish over time, while sex and type of salmon is necessary to determine whether mercury concentrations are consistent among these variables. Secondly, an analysis of where the salmon are located is needed. Salmon samples should be obtained from different depths of the lake and near and away from cities. Once multiple salmon are caught in each category and location, they can be placed on ice and shipped to the lab for sample preparation and analysis.

(b) To tackle this problem, a visit to the farm is necessary to examine the field and the surrounding environment, to observe the habits of the chickens, and to ask specific questions of the farmer. For example, a map of the property is essential. A feeding schedule and typical times when the chickens are outside would help to narrow potential sampling times. An understanding and access to manure collection in the building where chickens live and eat is necessary.

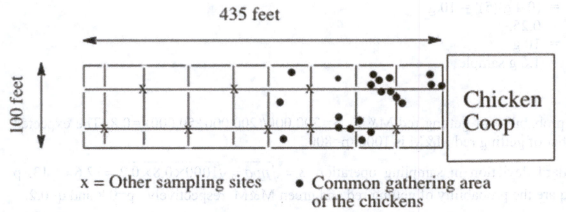

Figure 1. Roxarsone

After questions are answered and the chickens are observed to see if there are any common waste elimination locations, the sampling plan may be developed. A possible sampling grid is shown in Figure 2.

Figure 2. Potential Sampling Sites for Roxarsone Problem.

In areas where the chickens commonly gather, multiple grab samples on the surface and subsurface, 6 inches below, should be taken with a plastic trowel and placed in carefully labeled plastic bags. In areas less populated by the chickens, surface and subsurface samples could be taken every 30 feet. Samples from different depths of the manure pile that results from waste in the chicken coop building will also be needed. Three to four soil samples from the farm, away from the chickens coop or surrounding area, need to be taken to provide a blank for the farm soil.

With respect to sampling times, it would be best to sample shortly after waste elimination, at a time when the chickens are not outside and at least once after it rains to examine any runoff of the arsenic containing species. The sampling frequency would depend on variables such as how often the arsenic containing compounds are fed to the chickens and whether a specific set of chickens are followed from birth to slaughter.

In addition to sampling waste products and the surrounding soil, multiple chickens would need to be analyzed for arsenic. Finally, grab samples of several different bags of the feed additive would also need to be taken to obtain an accurate value of arsenic that is present in the feed additive. Future work at this site could also include sampling any ponds, rivers or lakes near the property for roxarsone and arsenic degradation products. A study of the rate of degradation for the roxarsone could also be done with the resulting samples.

CHAPTER 4   GOOD LABORATORY PRACTICE: QUALITY ASSURANCE AND METHOD VALIDATION

1. Good Laboratory Practice is the general principle of assuring quality operation of a laboratory, from management practices, laboratory personnel, method validation and quality control, to reporting and record keeping, in order to assure correctness of results produced in the laboratory.

2. QUA is the Quality Assurance Unit, who is responsible for implementing and assessing quality procedures, and Standard Operating Procedures (SOPs) that provide details for carrying out the laboratory operations.

3. See Question 2 above.

4. The QUA should be independent from the laboratory. It establishes the quality assurance and quality control procedures to be implemented, and monitors and   assesses them.

5. The problem is first defined, along with the data requirements. Then performance of the selected method must be validated to meet these requirements.

6. The minimum requirements of the method are decided, including accuracy and precision.

7. A technique refers to the technology to be used for a measurement, e.g., spectrophotometry. A method is the application of the technique, developing the proper chemistry or procedure for selective measurement. A procedure is the written directions for using the method. A protocol is a set of specifically prescribed directions that must be followed for official acceptance of the results.

8. Method validation generally requires studies to determine and validate selectivity, linearity, accuracy, precisision, sensitivity, range, limit of detection, limit of quantitation, and ruggedness or robustness.

9. The Response Factor is a way of assessing linearity, by determining if the response (with y-intercept subtracted) per unit concentration remains reasonably constant over the concentration ranges.

10. Besides the Response Factor (Question 9), the coefficient of determination (r^2), and a small y-intercept are measures of linearity range.

11. Accuracy is determined by recovery studies, comparing results with those of another method of known accuracy, or by analyzing a reference material. The ultimate measure of accuracy is from analysis of a standard reference material.

12. At least seven measurements (six degrees of freedom) should be made.

13. Repeatability = short term intralaboratory precision
    Ruggedness = long term intralaboratory precision
    Robustness (repeatability) = sensitivity to small changes in parameters
    Reproducibility (transferability) = interlaboratory precision or bias

14. Electronic records need to be backed up, archived, and recoverable. The data have to be secure, and any changes documented, with retention of original data. They must be transferable if the software is changed. There must be time- and date-stamped audit trails that can't be changed. Electronic signatures require both a username and password that are unique, not reassignable. The password should be changed periodically.

15. QA is the ongoing checking of the performance of a method. It includes appropriate quality control procedures, which provide quantitative measures of performance.

16. Quality control activities include maintaining control charts, using blind and random reference samples, and proficiency testing via collaborative laboratory studies.

17. A z-score is a measure of how close a laboratory's results in a collaborative study are to the known concentration, by how many standard deviations of the accepted concentration it differs.

18. Accreditation is when an authoritative body certifies a laboratory is competent to perform specific tasks.

19. See website for spreadsheets.

20. A. $C = 1$ mg/kg $= 10^{-6}$
    $S_R = 0.02C^{0.85}$
        $= 0.02(10^{-6})^{0.85}$
    Spreadsheet cell: $= .02(10^{\wedge}-6)^{\wedge}.85$
                    $= 1.55886e-07$
    $S_R = 1.6 \times 10^{-7}$
    %rsd $= (1.6 \times 10^{-7})/(10^{-6}) \times 100 = 16\%$

    B. %rsd $= 2C^{-0.15}$
    Spreadsheet cell: $= 2(10^{\wedge}-6)^{\wedge}-.15 = 15.88656469 = 16\%$

21. $z = (9.8 - 10.3)/0.5 = -1.0$. Your results are within one standard deviation of the accepted value.

CHAPTER 5 STIOCHIOMETRIC CALCULATIONS: THE WORKHORSE OF THE ANALYST

1. The first unit in each sample denotes the analyte unit, and the second denotes the sample unit. So the volume or the weight of either or both may be measured.

2. ppm = $\mu g/g$ or mg/kg (wt/wt) = $g/g \times 10^6$
   $\quad\quad$ = $\mu g/mL$ or mg/L (wt/vol) = $g/ml \times 10^6$
   $\quad\quad$ = ng/mL or $\mu L/L$ (vol/vol) = $mL/mL \times 10^6$
   ppb = ng/g or $\mu g/kg$ (wt/wt) = $g/g \times 10^9$
   $\quad\quad$ = ng/mL or $\mu g/L$ (wt/vol) = $g/mL \times 10^9$
   $\quad\quad$ = pL/mL or nL/L (vol/vol) = $mL/mL \times 10^9$

3. Eq. wt. = f.w./charge. This concept is used by physicians to give an overall view of the electrolyte balance.

4. A titration reaction should be stoichiometric and rapid, specific with no side reactions, quantitative, and there should be a marked change in a property of the solution at the equivalence point (when the reaction is complete). The four classes of titration are: acid-base, reduction-oxidation, precipitation, and complexometric.

5. The equivalence point of a titration is the point at which the reaction is complete, and the end point is the point at which it is observed to be complete.

6. A standard solution is one whose concentration is known to the degree of accuracy required in an analysis (e.g., titration). It is prepared by dissolving a known amount of sufficiently pure reagent (e.g., a primary standard) in a known volume of solvent, or else by titrating a known quantity of a pure reagent (primary standard) with an approximately prepared solution to standardize it.

7. A primary standard should be $\geq 99.98\%$ pure, be stable to drying temperatures, have a high formula weight, and possess the properties required for a titration.

8. So that a sufficiently large amount of it will have to be weighed for the titration that the error in weighing is small.

9. (a) 5.00 g/100 mL x (250)/100 = 12.5 g

   (b) 100g/100 mL x (500)/(100) = 5.00 g

   (c) 10.0 g/100 mL x (1000)/(100) = 100 g

10. (a) (52.3 g/1000 mL) x 100mL = 5.23 g/100 mL = 5.23% (wt/vol)

(b) (275 g/500 mL) x 100 mL = 55.0 g/100 mL = 55.0% (wt/vol)

(c) (3.65 g/200 mL) x 100 mL = 1.82 g/100mL = 1.82% (wt/vol)

11. (a) 244.27 (b) 218.16 (c) 431.73 (d) 310.18

12. (a) 500 mg/253 mg/mmol = 1.98 mmol $BaCrO_4$

(b) 500 mg/119 mg/mmol = 4.20 mmol $CHCl_3$

(c) 500 mg/389 mg/mmol = 1.28 mmol $KIO_3.HIO_3$

(d) 500 mg/137 mg/mmol = 3.65 mmol $MgNH_4PO_4$

(e) 500 mg/223 mg/mmol = 2.24 mmol $Mg_2P_2O_7$

(f) 500 mg/382 mg/mmol = 1.31 mmol $FeSO_4.C_2H_4(NH_3)_2SO_4.4H_2O$

13. 0.200 mol/L x 0.100 L = 0.0200 mol of each substance required

(a) 253 g/mol x 0.0200 mol = 5.06 g $BaCrO_4$

(b) 119 g/mol x 0.0200 mol = 2.38 g $CHCl_3$

(c) 390 g/mol x 0.0200 mol = 7.80 g $KIO_3.HIO_3$

(d) 137 g/mol x 0.0200 mol = 2.74 g $MgNH_4PO_4$

(e) 223 g/mol x 0.0200 mol = 4.46 g $Mg_2P_2O_7$

(f) 382 g/mol x 0.0200 mol = 7.64 g $FeSO_4.C_2H_4(NH_3)_2SO_4.4H_2O$

14. (a) mg NaCl = 1.00 mmol/mL x 1000 mL x 58.4 mg/mmol = 5.84×10^4 mg

(b) mg sucrose = 0.200 mmol/mL x 500 mL x 342 mg/mmol = 3.42×10^4 mg

(c) mg sucrose = 0.500 mmol/mL x 10.0 mL x 342 mg/mmol = 1.71×10^3 mg

(d) mg Na_2SO_4 = 0.200 mmol/mL x 10.0 mL x 142 mg/mmol = 284 mg

(e) mg KOH = 0.500 mmol/mL x 250 mL x 56.1 mg/mmol = 7.01×10^3 mg

(f) mg NaCl = 0.900 g/100 mL x 250 mL x 1000 mg/g = 2.25×10^3 mg

15. (a) $mL_{HCl} = 50.0 \text{ mmol}/(0.100 \text{ mmol/mL}) = 500 \text{ mL}$

(b) $mL_{NaOH} = 10.0 \text{ mmol}/(0.0200 \text{ mmol/mL}) = 500 \text{ mL}$

(c) $mL_{KOH} = 100/(0.0500 \text{ mmol/mL}) = 2.00 \times 10^3 \text{ mL}$

(d) $mL_{HBr} = 5.00 \text{ g}/(10.0 \text{ g}/100 \text{ mL}) = 50.0 \text{ mL}$

(e) $mL_{Na2CO3} = 4.00 \text{ g}/(5.00 \text{ g}/100 \text{ mL}) = 80.0 \text{ mL}$

(f) $1.00 \text{ mol HBr} = 80.9 \text{ g}$
$mL_{HBr} = 80.9 \text{ g}/(10.0 \text{ g}/100 \text{mL}) = 809 \text{ mL}$

(g) $0.500 \text{ mol Na}_2CO_3 = 0.500 \text{ g} \times 106.0 \text{ g/mol} = 53.0 \text{ g}$
$mL_{Na2CO3} = 53.0 \text{ g}/(5.00 \text{ g}/100 \text{ mL}) = 1.06 \times 10^3 \text{ mL}$

16. $\text{mmol Mn}^{2+} = \text{mmol Mn(NO}_3)_2 = 0.100 \text{ mmol/mL} \times 10.0 \text{ mL} = 1.00 \text{ mmol}$
$M_{Mn2+} + 1.00 \text{ mmol}/30.0 \text{ mL} = 0.0333 \text{ mmol/mL}$
$\text{mmol NO}_3^- = \text{mmol KNO}_3 + 2 \times \text{mmol Mn(NO}_3)_2$
$\qquad = 0.100 \text{ mmol/mL} \times 10.0 \text{ mL} + 2 \times 0.100 \text{ mmol/mL} \times 10.0 \text{ mL} = 3.00 \text{ mmol}$
$M_{NO3^-} = 3.00 \text{ mmol}/30.0 \text{ mL} = 0.100 \text{ mmol/mL}$
$\text{mmol K}^+ = \text{mmol KNO}_3 + 2 \times \text{mmol K}_2SO_4$
$\qquad = 0.100 \text{ mmol/mL} \times 10.0 \text{ mL} + 2 \times 0.100 \text{ mmol/mL} \times 10.0 \text{ mL} = 3.00 \text{ mmol}$
$M_{K+} = 3.00 \text{ mmol}/30.0 \text{ mL} = 0.100 \text{ mmol/mL}$
$\text{mmol SO}_4^{2-} = \text{mmol K}_2SO_4 = 0.100 \text{ mmol/mL} \times 10.0 \text{ mL} = 1.00 \text{ mmol}$
$M_{SO42-} = 1.00 \text{ mmol}/30.0 \text{ mL} = 0.0333 \text{ mmol/mL}$

17. $10.0 \text{ mmol/L} = 0.0100 \text{ mmol/mL}$
$0.0100 \text{ mmol/mL} \times 0.147 \text{ g/mmol} = 0.00147 \text{ g CaCl}_2.2H_2O/\text{mL}$

18. (a) $(10.0 \text{ g}/250 \text{ mL})/(98.1 \text{ g/mol}) \times 1000 \text{ mL/L} = 0.408 \, M \, H_2SO_4$

(b) $(6.00 \text{ g}/500 \text{ mL}) /(40.0 \text{ g/mol}) \times 1000 \text{ mL/L} = 0.300 \, M \, NaOH$

(c) $(25.0 \text{ g/L})/(170 \text{ g/mol}) = 0.147 \, M \, AgNO_3$

19. (a) $(0.100 \text{ mol/L})(142 \text{ g/mol})(0.500 \text{ L}) = 7.10 \text{ g Na}_2SO_4$

(b) $(0.250 \text{ mol/L})(392 \text{ g/mol})(0.500 \text{ L}) = 49.0 \text{ g Fe(NH}_4)_2(SO_4)_2.6H_2O$

(c) $(0.667 \text{ mol/L})(328 \text{ g/mol})(0.500 \text{ L}) = 109 \text{ g Ca(C}_9H_6ON)_2$

20. (a) $(250 \text{ mL})(0.100 \text{ mmol/mL})(56.1 \text{ mg/mmol})(1.00 \times 10^{-3} \text{ g/mg}) = 1.40 \text{ g KOH}$

(b) $(1.00 \text{ L})(0.0275 \text{ mol/L})(294 \text{ g/mol}) = 8.08 \text{ g K}_2Cr_2O_7$

(c) $(500 \text{ mL})(0.0500 \text{ mmol/mL})(160 \text{ mg/mmol})(1.00 \times 10^{-3} \text{ g/mg}) = 4.00 \text{ g CuSO}_4$

21. $[(0.380 \text{ g/g} \times 1.19 \text{ g/mL})/36.5 \text{ g/mol}] \times 1000 \text{ mL/L} = 12.4 \text{ mol/L}$ in stock solution
    $12.4 \text{ mmol/mL} \times X \text{ mL} = 0.100 \text{ mmol/mL} \times 1000 \text{ mL}$
    $X = 8.06 \text{ mL}$ must be diluted

22. (a) $[(0.700 \text{ g/g} \times 1.668 \text{ g/mL})/100.5 \text{ g/mol}] \times 1000 \text{ mL/L} = 11.6 \text{ mol/L}$

    (b) $[(0.690 \text{ g/g} \times 1.409 \text{ g/mL})/63.01 \text{ g/mol}] \times 1000 \text{ mL/L} = 15.4 \text{ mol/L}$

    (c) $[(0.850 \text{ g/g} \times 1.689 \text{ g/mL})/989.0 \text{ g/mol}] \times 1000 \text{ mL/L} = 14.6 \text{ mol/L}$

    (d) $[(0.995 \text{ g/g} \times 1.051 \text{ g/mL})/60.05 \text{ g/ml}] \times 1000 \text{ mL/L} = 17.4 \text{ mol/L}$

    (e) $[(0.280 \text{ g/g} \times 0.898 \text{ g/mL})/17.03 \text{ g/mol}] \times 1000 \text{ mL/L} = 14.8 \text{ mol/L}$

23. $6.0 \times 10^{-6} \text{ mol/250 mL} = 24 \times 10^{-6} \text{ mol/L Na}_2\text{SO}_4$
    $= 48 \times 10^{-6} \text{ mol/L of Na}^+$ and $24 \times 10^{-6} \text{ mol/L of SO}_4^{2-}$
    $48 \times 10^{-6} \text{ mol/L} \times 23 \times 10^3 \text{ mg/mol} = 1.1 \text{ mg/L Na}^+$
    $24 \times 10^{-6} \text{ mol/L} \times 96 \times 10^3 \text{ mg/mol} = 2.3 \text{ mg/L SO}_4^{2-}$

24. $325 \text{ mg/L} \times 0.100 \text{ L} = 32.5 \text{ mg K}^+$
    $(32.5 \text{ X } 10^{-3} \text{ g})/(39.1 \text{ g/mol}) = 8.31 \times 10^{-4} \text{ mol K}^+ = 8.31 \times 10^{-4} \text{ mol (C}_6\text{H}_5)_4\text{B}^-$
    $8.31 \times 10^{-4} \text{ mol} \times 319 \text{ g/mol} \times (1000 \text{ mL/L})/(250 \text{ mL}) \times 10^3 \text{ mg/g}$
    $= 1.06 \times 103 \text{ mg/L (C}_6\text{H}_5)_4\text{B}^-$

25. $\text{g/mL} = \text{mg/L} \times 10^{-6} = 1.00 \times 10^{-6} \text{ g/mL} = 1.00 \times 10^{-3} \text{ g/L}$

    (a) $(1.00 \times 10^{-3} \text{ g/L})/170 \text{ g/mol}) = 5.88 \times 10^{-6} \text{ mol/L AgNO}_{3\backslash}$

    (b) $(1.00 \times 10^{-3} \text{ g/L})/(343 \text{ g/mol}) = 2.92 \times 10^{-6} \text{ mol/L Al}_2(\text{SO}_4)_3$

    (c) $(1.00 \times 10^{-3} \text{ g/L})/(44.0 \text{ g/mol}) = 2.27 \times 10^{-5} \text{ mol/L CO}_2$

    (d) $(1.00 \times 10^{-3} \text{ g/L})/(633 \text{ g/mol}) = 1.58 \times 10^{-6} \text{ mol/L (NH}_4)_4\text{Ce(SO}_4)_4.2\text{H}_2\text{O}$

    (e) $1.00 \times 10^{-3} \text{ g/L})/(36.5 \text{ g/mol}) = 2.73 \times 10^{-5} \text{ mol/L HCl}$

    (f) $(1.00 \times 10^{-3} \text{ g/L})/(100 \text{ g/mol}) = 1.00 \times 10^{-5} \text{ mol/L HClO}_4$

26. $2.50 \times 10^{-4} \text{ mol/L} \times 103 \text{ mmol/mol} = 0.250 \text{ mmol/L}$

    (a) $0.250 \text{ mmol/L} \times 40.1 \text{ mg/mmol} = 10.0 \text{ mg/L Ca}^{2+}$

    (b) $0.250 \text{ mmol/L} \times 111 \text{ mg/mmol} = 27.8 \text{ mg/L CaCl}_2$

(c) 0.250 mmol/L x 63.0 mg/mmol $= 15.8$ mg/L HNO_3

(d) 0.250 mmol/L x 65.1 mg/mmol $= 16.3$ mg/L KCN

(e) 0.250 mmol/L x 54.9 mg/mmol $= 13.7$ mg/L Mn^{2+}

(f) 0.250 mmol/L x 119 mg/mmol $= 29.8$ mg/L MnO_4^-

27. 1.00 mg/L $Fe^{2+} = 0.00100$ g/L
    $(0.00100$ g/L$)/(55.8$ g/mol$) = 1.79$ x 10^{-5} mol/L
    1 mol $FeSO_4.(NH_4)_2SO_4.6H_2O$ contains 1 mol Fe
    Therefore, require 1.79 x 10^{-5} mol of this.
    1.79 x 10-5 mol x 392 g/mol $= 0.00702$ g $FeSO_4.(NH_4)_2SO_4.6H_2O$

    It is simpler to multiply the weight by the ratio of the formula weights:
    g $FeSO_4.(NH_4)_2SO_4.6H_2O$ = g Fe x [f.w. $FeSO_4.(NH_4)_2SO_4.6H_2O$/at. wt. Fe]
    $= 0.001$ x $(392)/(55.8) = 0.00702$ g

    The number of iron atoms in the numerator and denominator of the formula   weight ratio
    must be the same.
    $(0.00702$ g/L$)(/392$ g/mol$) = 1.79$ x 10^{-5} mol/L

28. (a) % $Cr_2O_3 = [0.560$ mg$/456$ mg$]$ x $100\% = 0.123\%$

    (b) ppt $Cr_2O_3 = [0.560$ mg$/456$ mg$]$ x 1000 ‰

    (c) ppm $Cr_2O_3 = [0.560$ mg$/456$ mg$]$ x $10^6 = 1.23$ x 10^3 ppm

29. 100 mg/L x $10^{-6} = 1.00$ x 10^{-4} g/mL $= 0.100$ g/L

    (a) $(0.100$ g/L$)/(23.0$ g/mol$) = 4.35$ x 10^{-3} mol/L Na^+
        The molar concentration of NaCl is the same as that of Na^+ (and Cl^-)
        4.35 x 10^{-3} mol/L x 58.4 g/mol $= 0.254$ g/L NaCl

    (b) $(0.100$ g/L$)/(35.5$ g/mol$) = 2.82$ x 10-3 mol/L $Cl^- =$ NaCl
        2.82 x 10^{-3} mol/L x 58.4 g/mol $= 0.165$ g/L NaCl

30. 250 mg/L x $10^{-6} = 2.50$ x 10^{-4} g/mL $= 0.250$ g/L K^+
    $(0.250$ g/L$)/(39.1$ g/mol$) = 6.39$ x 10^{-3} mol/L $K^+ =$ KCl $=$ Cl^-

    The millimoles before and after dilution are equal. So,
    6.39 x 10^{-3} mmol/mL x X mL $= 1.00$ x 10^{-3} mmol/mL x 1000 mL
    X $= 156$ mL KCl required

31. 500 mg/L x $10^{-6} = 5.00$ x 10^{-4} g/mL $= 0.500$ g/L $KClO_3$. There is 1 $K^+/KClO_3$.
    0.500 g $KClO_3$ x (K/ $KClO_3$) $= 0.500$ g x $(39.1)/(122) = 0.160$ g K^+

32. 12.5 mL x X M = 500 mL x 0.1225 M
    X = 5.00 M

33. Let x = mL 0.50 M H_2SO_4
    0.35 M x (65 + x) mL = 0.20 M x 65 mL + 0.50 M x x mL
    x = 65 mL

34. mmol NaOH = 50 x 0.10 = 5.0 mol
    mL H_2SO_4 to neutralize NaOH = 5.0/(0.10 x 2) = 25 mL
    Let x = additonal mL H_2SO_4 required
    0.050 M x (50 + 25 + x) mL = 0.10 M x x mL
    x = 75 mL
    Therefore, total volume = 25 + 75 = 100 mL

35. The most concentrated solution, 1.00 x 10^{-4} M, would require 1:1,000 dilution of the stock
    solution, and you can not do this with the glassware provided. Prepare a diluted stock
    solution. The most dilute possible is 1:100 (1 mL diluted to 100 mL):
    0.100 M x 1.00 mL = M_A x 100 mL
    M_A = 1.00 x 10^{-3} M

    This can be diluted appropriately to give the desired concentrtions.
    1.00 x 10^{-5} M: dilute 1 mL to 100 mL (1:100)
    2.00 x 10^{-5} M: dilute 2 mL to 100 mL (1:50)
    5.00 x 10^{-5} M: dilute 5 mL to 100 mL (1:20)
    1.00 x 10^{-4} M: dilute 10 mL to 100 mL (1:10)

36. The dilution factor for the original solution is 1:5 (50 mL aliquot diluted to 250 mL).
    Hence, the concentration in the original solution is 5 x (1.25 x 10^{-5} M) = 6.25 x 10^{-5} M.
    We have 250 mL at this concentration.
    $mmol_{Mn}$ = 6.25 x 10^{-5} M x 250 mL = 0.0156 mmol
    mg_{Mn} = 0.0156 mmol x 54.9 mg/mmol = 0.858 mg = 8.58 x 10^{-4} g
    % Mn = [(8.58 x 10^{-4} g)/(0.500 g)] x 100% = 0.172%

37. Stock solution concentration (analyte): <u>4.809 x 10^{-3} M</u>
    Stock solution concentration (internal standard): <u>5.962 x 10^{-3} M</u>

Concentration Cu (µM)	Volume of analyte stock (mL)	Volume of internal standard stock (mL)	Volume of diluent (mL)	Total Volume (mL)
10.00	0.02080	0.04193	9.937	10.000
25.00	0.05199	0.04193	9.906	10.000
50.00	0.1040	0.04193	9.854	10.000
100.0	0.2080	0.04193	9.750	10.000
200.0	0.4159	0.04193	9.542	10.000

Volume of analyte stock (for 10 μM standard):
$C_1V_1 = C_2V_2$
$(4.809 \times 10^{-3} \text{ M}) (x \text{ mL}) = (1.000 \times 10^{-5} \text{ M}) (10.000 \text{ mL})$
$\underline{x = 0.02080 \text{ mL}}$

Volume of internal standard (25 μM in all solutions):
$C_1V_1 = C_2V_2$
$(5.962 \times 10^{-3} \text{ M}) (x \text{ mL}) = (2.500 \times 10^{-5} \text{ M}) (10.000 \text{ mL})$
$\underline{x = 0.04193 \text{ mL}}$

Volume of diluent = Total volume – Vol of analyte stock – Vol of internal standard:
$x = 10.000 - 0.02080 - 0.04193$
$\underline{x = 9.937 \text{ mL}}$

38. PFP

    (a) Based on dilution factor, $C_1V_1 = C_2V_2$, Therefore the concentrations in the brewed coffee
        are 938, 798, 824 and 900 mg/L.
        The caffeine concentration in the brewed coffee is 865 ± 65 mg/L or 865 ± 65 ppm

    (b) See for example http://coffeetea.about.com/library/blcaffeine.htm
        A serving of regular coffee contains caffeine of (60 -120) mg/8 oz, i.e., 0.264 – 0.529
        mg/mL brewed coffee.
        Decaffeinated coffee has 1-5 mg caffeine/8 oz, i.e., 4.4 -22 μg/mL.
        Tom is not only drinking regular coffee, he has made his brew extra strong not
        decaffeinated coffee.

    (c) Tom's daily caffeine intake is 3 cup × 8 oz/cup × 28.35 mL/oz ×865 mg/L=588.5 mg
        Tom's caffeine intake is not in the known no-effects zone.

39. The neutralization reaction is $Na_2CO_3 + H_2SO_4 \rightarrow Na_2SO_4 + 2H_2O$

$$\therefore \%Na_2CO_3 = \frac{M\,H_2SO_4 \times mL\,H_2SO_4 \times 1(\text{mmol } H_2SO_4/\text{mmol } Na_2CO_3) \times \text{f.w. } Na_2CO_3}{\text{mg sample}} \times 100\%$$

$$\therefore 96.8\% = \frac{M\,H_2SO_4 \times 36.8 \times 1 \times 106.0}{678} \times 100\%$$

$M_{H2SO4} = 0.171 \, M$

40. 40 mL x 0.10 M = 4.0 mmol NaOH = mmol sulfamic acid
    4.0 mmol = (x mg)/97 mg/mmol; x = 390 mg sulfamic acid

41. % H_3A = {[0.1087 M x 38.31 mL x 1/3 (mmol H_3A/mmol NaOH)
    x 192.1 mg/mmol]/(267.8 mg)} x 100% = 99.57%

42. mmol Ca^{2+} = mmol EDTA

mg Ca^{2+} = 1.87 x 10^{-4} M x 2.47 mL x 40.1 mg/mmol = 0.0185 mg

(200 μL)/(1000 μL/mL) = 0.200 mL

(0.0185 mg/0.200 mL) x 100 mL/dL = 9.25 mg/dL

43. (a) % Cl = {[0.100 M x 27.2 mL x 1 (mmol Cl^-/mmol Ag^+)

x 35.5 mg/mmol]/372 mg)} x100% = 26.0%

(b) % $BaCl_2.2H_2O$ = {[0.100 M x 27.2 mL x ½ (mmol $BaCl_2$/mmol Ag^+)

x 244 mg/mmol]/372 mg} x 100% = 89.2%

44. Each $Cr_2O_7^{2-}$ reacts with $6Fe^{2+}$ (= $3Fe_2O_3$), therefore, % Fe_2O_3 =

[($M_{Cr2O72-}$ - x $mL_{Cr2O72-}$ - 3(mmol Fe_2O_3/mmol $Cr_2O_7^{2-}$) x f.w.$_{Fe2O3}$ (mg/mmol)/(mg sample)] x

100% = [(0.0150 x 35.6 x 3 x 160)/(1680)] x 100%

= 15.3%

45. 1 Ca = 1 $H_2C_2O_4$ = 2/5 MnO_4^-

Therefore, 1 MnO_4^- = 5/2 CaO

% CaO = [(35.6 mL x 0.0200 mmol/mL x 5/2 x 56.1 mg/mmol)/(2000 mg)] x 100%

= 4.99%

46. M_{MnO4-} = (4680 mg)/($KMnO_4$ x 500 mL) = (4680 mg)/(159.0 mg/mmol) x 500mL)

= 0.0592 mmol/mL

1 MnO_4^- = 2.5 Fe_2O_3

∴ 35.6% = [(0.0592 mmol/mL x mL x 2.5 x Fe_2O_3 mg/mmol)/(500 mg)] x 100%

= [(0.0592 x mL x 2.5 x 159.7)/(500 mg)] x 100%

mL = 7.53

47. Let a = % $BaCl_2$ = mL $AgNO_3$

1 Ag = ½ $BaCl_2$

∴ a = [(0.100 M x a mL x ½ x $BaCl_2$)/(mg)] x 100%

1 = [(0.100 x ½ x 208)/(mg)] x 100%

mg = 1.04 x 10^3 mg sample

48. mg = 1.04 x 10^3 mg sample

x 133 mg/mmol]/(250 mg)} x 100% = 86.2%

mmol Al in 350 mg = [(350 mg x 0.862)/133 (mg $AlCl_3$/mol)]

= 2.27 mmol Al (= mmol $AlCl_3$)

2.27 mmol = 0.100 x x mL

x = 22.7 mL EDTA

49. 100% = [(0.1027 M x 28.78 mL x f.w.$_{HA}$)/(425.2 mg)] x 100%

f.w.$_{HA}$ = 143.9

50. (462 mg AgCl)/143 mg/mmol) = 3.23 mmol AgCl = mmol HCl
    M_{HCl} x 25.0 mL = 3.23 mmol
    M_{HCl} = 0.129 M
    % $Zn(OH)_2$ = {[0.129 M x 37.8 mL x ½ (mmol $Zn(OH)_2$/mmol HCl)
    x 99.4 mg/mmol]/(287 mg)} x 100% = 88.4%

51. mmol $KHC_2O_4.H_2C_2O_4.2H_2O$ = 1/3 mmol NaOH = 1/3 x 46.2 mL x 0.100 mmol/mL
    = 1.54 mmol
    mmol $C_2O_4^{2-}$ = 2 x mmol $KHC_2O_4.H_2C_2O_4.2H_2O$ = 2 x 1.54 = 3.08 mmol
    Each $C_2O_4^{2-}$ = 2/5 MnO_4^-
    Therefore, mmol $KMnO_4$ =2/5 x 3.08 = 1.23 mmol
    0.100 mol/mL x mL_{MnO4-} = 1.23 mmol
    mL_{MnO4-} = 12.3 mL

52. mmol Na_2CO_3 = ½ mmol HCl reacted
    ∴ % Na_2CO_3 = {[(0.0100 x 50.0 – 0.100 x 5.6) x ½ x 106.0]/(500 mg)} x 100%
    = 47.1%

53. 1 MnO_4^- = 5/2 H_2O_2; 1 Fe^{2+} = 1/5 MnO_4^-
    Therefore, mmol H_2O_2 = mmol MnO_4^- reacted x 5/2
    = (mmol MnO_4^- taken - mmol MnO_4^- unreacted) x 5/2
    = (mmol MnO_4^- taken – mmol Fe^{2+} x 1/5) x 2/5
    Therefore, % H_2O_2 =
    {[(M_{MnO4-} x mL_{MnO4-} - M_{Fe2+} x mL_{Fe2+} x 1/5) x 5/2 x f.w.$_{H2O2}$]/(mg_{sample})} x 100%
    = {[(25.0 x 0.0215 – 5.10 x 0.112 x 1/5) x 5/2 x34.0]/(587)} x 100% = 6.13%

54. mmol excess I_2 = ½ mmol $S_2O_3^{2-}$
    mmol H_2S = mmol reacted I_2
    ∴ mg S = (0.00500 M x 10.0 mL – ½ x 0.00200 M x 2.6 mL) x 32.06 = 1.52 mg

55. The reaction is Ba^{2+} + H_2EDTA^{2-} = $Ba-EDTA^{2-}$ + $2H^+$
    Therefore, mmol EDTA = mmol BaO
    0.100 mmol/mL x 1 mL = (mg BaO/(BaO) = (mg BaO)/(153 mg/mmol)
    Titer = 15.3 mg BaO per milliliter EDTA

56. 1 Fe reacts with 1/5 MnO_4^-
    Therefore, 5/2 mmol MnO_4^- = mmol Fe_2O_3
    5/2 x 0.0500 mmol/mL x 1 mL = (mg Fe_2O_3)/(Fe_2O_3) = (mg Fe_2O_3)/(159.7 mg/mmol)
    Titer = 20.0 mg Fe_2O_3 per milliliter of $KMnO_4$

57. mmol Ag^+ = mmol X^-
    M x 1 mL = (22.7 mg)/(Cl) = (22.7 mg)/(35.4 mg/mmol)
    M_{AgNO3} = 0.641 M
    0.641 M x 1 mL = (mg Br)/(Br) = (mg Br)/(79.9 mg/mmol)
    Titer = 51.2 mg Br/mL
    Or, titer = 22.7 mg x (Br/Cl) = 22.7 mg x (79.9/35.4) = 51.2 mg Br/mL

58. (a) (36.46 g/mol)/(1 eq/mol) = 36.46 g/eq

   (b) (171.34 g/mol)/2 eq/mol) = 85.67 g/eq

   (c) (389.91 g/mol)/(1eq/mol) = 389.91 g/eq

   (d) (82.08 g/mol)/(2 eq/mol) = 41.04 g/eq

   (d) (60.05 g/mol)/(1 eq/mol) = 60.05 g/eq

59. (a) (0.250 eq/L)/(1 mol/eq) = 0.250 mol/L

   (b) (0.250 eq/L)/(1/2 mol/eq) =0.125 mol/L

   (c) (0.250 eq/L)/(1 mol/eq) = 0.250 mol/L

   (d) (0.250 eq/L)/(1/2 mol/eq) = 0.125 mol/L

   (e) (0.250 eq/L)/(1 mol/eq) = 0.250 mol/L

60. (a) eq wt = (128.1 g/mol)/(1 eq/mol) = 128.1 g/eq

   (b) Each C is oxidized from +3 to +4, so the change is 2 electrons/$HC_2O_4^-$.
       eq wt = (128.1 g/mol)/(2 eq/mol) = 64.05 g/eq

61. eq wt = (216 g/mol)/(2 eq/mol) = 108.3 g/eq

62. (a) 151.91/1 = 151.91 g/eq

   (b) 34.08/2 = 17.04 g/eq

   (c) 34.01/2 = 17.00 g/eq

   (e) 34.01/2 = 17.00 g/eq

63. meq = (mg/[eq wt (mg/meq)]) = 4.093 meq

64. 7.82 g NaOH/(40.0 g/eq) = 0.196 eq
    9.26 g $Ba(OH)_2$/(171/2 g/eq) = 0.108 eq
    N = [(0.196 + 0.108) eq/(500 mL)] x 1000 mL/L = 0.608 eq/L

65. Each As undergoes 2 electron change. Therefore,
    eq wt = As_2O_3/4 = 197.8/4 = 49.45 g/eq
    0.1000 N = 0.1000 eq/L; 0.1000 eq/L x 49.45 g/eq = 4.945 g/L

66. .2.73 g x 0.980 = 2.68 g $KHC_2O_4.H_2C_2O_4$
    2.68 g/(218/3 g/eq) = 0.0369 eq
    1.68 $KHC_8H_4O_4$/(204 g/eq) = 0.00823 eq
    N = [(0.0369 + 0.00823) eq/(250 mL)] x 1000 mL/L = 0.180 eq/L

67. The carbon in each oxalate ($C_2O_4^{2-}$) is oxidized from +3 to +4, releasing 2 electrons/$C_2O_4^{2-}$. Therefore, the equivalent weight of $KHC_2O_4.H_2C_2O_4.2H_2O$ as a reducing agent is one-fourth its formula weight. Therefore, $N_{red} = N_{acid}$ x 4/3 = 0.200 x 4/3 = 0.267 N.

68. Assume 1.00 N as acid, therefore, 3.62 N as reducing agent. Acidity is due to $KHC_2O_4.H_2C_2O_4$, and so the concentration of this is 1.00 N as an acid or 1.00(4/3) = 1.33 N as a reducing agent. If we have 1 L of solution, the normality and equivalents are equal. Therefore,
    1.33 eq $KHC_2O_4.H_2C_2O_4$ + x eq $Na_2C_2O_4$ = 3.62 eq (as reducing agent)
    x = 2.29 eq $Na_2C_2O_4$
    1.33 eq x (218 g/4) g/eq = 72.5 g $KHC_2O_4.H_2C_2O_4$
    2.29 eq x (134/2) g/eq = 153 g $Na_2C_2O_4$
    ∴ ratio = 72.5/153 = 0.474 g $KHC_2O_4.H_2C_2O_4$ /g $Na_2C_2O_4$

69. meq = N x mL = 0.100 meq/mL x 1000 mL = 100.0 meq
    mg = meq x eq wt (mg/meq) = 100.0 meq x (294.2/6)(mg/meq)
    = 4903 mg (4.903 g)

70. (300 mg/dL)/(0.1 L/dL) = 3.00 x 10^3 mg/L
    (3.00 x 10^3 mg/L)/35.5 mg/meq) = 84.5 meq/L

71. [(5.00 meq/L) x (40.1/2 mg/meq)]/(10 dL/L) = 10.0 mg/dL

72. There are 150 mmol/L of NaCl. Therefore,
    150 mol/L x 0.0584 g/mmol = 8.76 g/L

73. g Mn = g Mn_3O_4 x (3Mn/Mn_3O_4) g Mn/g Mn_3O_4
    = 2.58 g Mn_3O_4 x [3(54.9)/228.8] = 1.85$_7$ g Mn

74. (a) g Zn = g $Zn_2Fe(CN)_6$ x [2Zn/$Zn_2Fe(CN)_6$] g Zn/g $Zn_2Fe(CN)_6$
    = 0.348 x [(2(65.4)/342.7] = 0.348 x 0.369 (g Zn/g $Zn_2Fe(CN)_6$) = 1.32$_8$ g Zn
    (0.369 is the gravimetric factor)

    (b) g $Zn_2Fe(CN)_6$ = g Zn x (1/2 $Zn_2Fe(CN)_6$/Zn) g $Zn_2Fe(CN)_6$/g Zn
    = 0.500 g Zn x [1/2(342.7/65.4)] = 1.31$_0$ g $Zn_2Fe(CN)_6$

75. 3Mn/Mn_3O_4 = = 3(54.938)/228.81 = 0.7203$_1$ g Mn/g Mn_3O_4
    3Mn_2O_3/2Mn_3O_4 = 3(157.88)/2(228.81) = 2.0700 g Mn_2O_3/g Mn_3O_4
    Ag_2S/$BaSO_4$ = 247.80/233.40 = 1.0617$_0$ g Ag_2S/g $BaSO_4$
    $CuCl_2$/2AgCl = 134.45/2(143.32) = 0.46906 g $CuCl_2$/g AgCl
    MgI_2/PbI_2 = 278.12/261.00 = 1.0656 g MgI_2/g PbI_2

76 Let

     x = Moles NaCl
     y = Moles KCl
     z = Moles AgCl

     x x 58.44 g + y x 74.55 g = 10.00 g
     x_l + y = z = 21.62 g/143.32 g = 0.1509
     y = 0.1509 – x

     x_l x 58.44 g + (0.1509 - x) x 74.55 g = 10.00 g
     x x (58.44 – 74.55) = 10.00 – 0.1509 x 74.55
     x = (10.00 – 0.1509 x 74.55)/(58.44 – 74.55) = 0.07734

     % NaCl = (0.07734 x 58.44)/10.00 x 100% = 45.20%

CHAPTER 6   GENERAL CONCEPTS OF CHEMICAL EQUILIBRIUM

1.   K = [C][D]/[A][B] = 2.0 x 10^3

At equlibrium,
[A] = x
[B] = (0.80 – 0.30) + x = 0.50 + x ≈ 0.50 M
[C] = [D] = 0.30 – x ≈ 0.30 M
[(0.30)(0.30)]/[(x)90.50)] = 2.0 x 10^3
x = 9.0 x 10^{-5} M = [A] (we were justified in neglecting x above)

2.   A + B = 2C
$[C]^2$/[A][B] = 5.0 x 10^6

The reaction is limited by the amount of A (0.40 M) available to react. At equilibrium:
[A] = x
[B] = (0.70 – 0.40) + x = 0.30 + x ≈ 0.30
[C] = 2(0.40) – x = 0.80 – x ≈ 0.80
$0.80)^2$/(x)(0.30) = 5.0 x 10^6
x = 4.3 x 10^{-7} M = [A]
See the text website for a video using Goal Seek to solve this problem.

3.   HA = H^+ + A^-
$[H^+][A^-]$/[HA] = K_{eq} = 1.0 x 10^{-3}

At equilibrium,
$[H^+]$ = $[A^-]$ = x
[HA] = 1.0 x 10^{-3} – x. Since [HA] < 100 K_{eq}, we can't neglect x. Solve for the quadratic formula.
(x)(x)/1.0 x 10^{-3} – x) = 1.0 x 10^{-3}
x^2 + 1.0 x 10^{-3}x – 1.0 x 10^{-6} = 0
$$x = \frac{(-1.0x10^{-3}) \pm \sqrt{(1.0x10^{-3})^2 + 4.0x10^{-6})}}{2} = 6.0x10^{-4} M$$
% dissociated = [(6.0 x 10^{-4})/(1.0 x 10^{-3})] x 100% = 60%

4.   HCN   = H^+ + CN^-
1.0 x 10^{-3} – x   x   x

Neglect x compared to 1.0 x 10^{-3} (C>>200 x K_{eq})
[H+][CN-]/[HCN] = 7.2 x 10^{-10}
(x)(x)/(1.0 x 10^{-3}) = 7.2 x 10^{-10}
x = 8.5 x 10^{-7} M
% dissociated = [(8.5 x 10^{-7})/(1.0 x 10^{-3})] x 100% = 0.085%

5. $HA = H^+ + A^-$

At equilibrium,
$[H^+] = x$
$[A^-] = 10 \times 10^{-2} + x \approx 1.0 \times 10^{-2}$
$[HA] = 1.0 \times 10^{-3} - x \approx 1.0 \times 10^{-3}$ (A^- will suppress the dissociation, so let's assume x is now small.)
$[H^+][A^-]/[HA] = 1.0 \times 10^{-3}$
$(x)(1.0 \times 10^{-2})/(1.0 \times 10^{-3}) = 1.0 \times 10^{-3}$
$x = 1.0 \times 10^{-4}\ M$ = concentration dissociated
% dissociated = $[(1.0 \times 10^{-4})/(1.0 \times 10^{-3})] \times 100\% = 10\%$

6. $H_2S = H^+ + HS^-$ $K_1 = [H^+][HS^-]/[H_2S] = 9.1 \times 10^{-8}$
$HS^- = H^+ + S^{2-}$ $K_2 = [H^+][S^{2-}]/[HS^-] = 1.2 \times 10^{-15}$

Overall:
$H_2S = 2H^+ + S^{2-}$ $K = [H^+]^2[S^{2-}]/[H_2S] = K_1K_2 = (9.1 \times 10^{-8})(1.2 \times 10^{-15})$
$= 1.1 \times 10^{-22}$

7. The initial analytical concentrations after mixing but before reaction, are
$[Fe^{2+}] = 0.06\ M$
$[Cr_2O_7^{2-}] = 0.01\ M$
$[H^+] = 1.14\ M$

There are stoichiometrically equal conentrations of Fe^{2+} and $Cr_2O_7^{2-}$. At equilibrium, 0.02 mol/L $Cr_2O_7^{2-}$ has reacted with 0.14 mol/L H^+, leaving 1.00 mol/L H^+ (plus the amount from the reverse equilibrium reaction, so

$6Fe^{2+}$	$+$	$Cr_2O_7^{2-}$	$+$	H^+	$=$	$6Fe^{3+}$	$+ 2Cr^{3+}$	$+7H_2O$
6x		x		1.00 + 14x		0.06 – 6x	0.02 – 2x	
				≈ 1.00		≈ 0.06	≈ 0.02	

$([Fe^{3+}]^6[Cr^{3+}]^2)/([Fe^{2+}]^6[Cr_2O_7^{2-}][H^+]^{14}) = 1 \times 10^{57}$
$[(0.06)^6(0.02)^2]/[(6x)^6(x)(1.00)^{14}] = 1 \times 10^{57}$
$x^7 = 4 \times 10^{-73} = 40{,}000 \times 10^{-77}$
$x = [Cr_2O_7^{2-}] = 5 \times 10^{-11}\ M$
(See Appendix B for a review of calculating odd roots)

$[Fe^{2+}] = 6x = 3 \times 10^{-10}\ M$

8. (a) Equilibria:
$Bi_2S_3 = 2Bi^{3+} + 3S^{2-}$
$S^{2-} + H^+ = HS^-$
$HS^- + H^+ = H_2S$
$H_2O = H^+ + OH^-$
$3[Bi^{3+}] + [H^+] = 2[S^{2-}] + [HS^-] + [OH^-]$

(b) Equilibria:

$Na_2S \rightarrow 2Na^+ + S^{2-}$

$S^{2-} + H^+ = HS^-$

$HS^- + H^+ = H_2S$

$H_2O = H^+ + OH^-$

$[Na^+] + [H^+] = 2[S^{2-}] + [HS^-] + [OH^-]$

9. Mass balance:

(Cd): $0.100 = [Cd^{2+}] + [Cd(NH_3)^{2+}] + [Cd(NH_3)_2^{2+}] + [Cd(NH_3)_3^{2+}] + [Cd(NH_3)_4^{2+}]$

(N): $0.400 = [NH_4^+] + [NH_3] + [Cd(NH_3)^{2+}] + [Cd(NH_3)_2^{2+}] + [Cd(NH_3)_3^{2+}]$
$+ [Cd(NH_3)_4^{2+}]$

(Cl): $0.200 = [Cl^-]$

Charge balance:

$2[Cd^{2+}] + [H^+] = 2[Cd(NH_3)^{2+}] + 2[Cd(NH_3)_2^{2+}] + 2[Cd(NH_3)_3^{2+}] + 2[Cd(NH_3)_4^{2+}]$
$= [Cl^-] + [OH^-]$

10. (a) Charge balance (C.B.): $[H^+] = [NO_2^-] + [OH^-]$. $\therefore$ $[NO_2^-] = [H^+] - [OH^-]$

(b) Mass balance (M.B.): $0.2 = [CH_3COO^-] + [CH_3COOH]$ (1)

C.B: $[H+] = [CH_3COO^-] + [OH^-]$ (2)

Combining Equations (1) and (2): $[CH_3COOH] = 0.2 - [CH_3COO^-]$
$= 0.2 - ([H^+] - [OH^-]) = 0.2 - [H^+] + [OH^-]$

(c) M.B.: $0.1 = [H_2C_2O_4] + [HC_2O_4^-] + [C_2O_4^{2-}]$ (3)

C.B.: $[H+] = [OH^-] + [HC_2O_4^-] + 2[C_2O_4^{2-}]$ (4)

Combining Equations (3) and (4): $[H_2C_2O_4] = 0.1 - ([HC_2O_4^-] + [C_2O_4^{2-}])$
$= 0.1 - ([H+] - [OH^-] - [C_2O_4^{2-}]) = 0.1 - [H^+] + [OH^-] + [C_2O_4^{2-}]$

(d) M.B.: $0.1 = [CN^-] + [HCN] = [K^+]$ (5)

C.B.: $[K+] + [H^+] = [OH^-] + [CN^-]$ (6)

Combining Equations (5) and (6):
$[HCN] = 0.1 - [CN^-] = 0.1 - (0.1 + [H^+] - [OH^-]) = [H^+] - [OH^-]$

(e) M.B.: $0.1 = [H_3PO_4] + [H_2PO_4^-] + [HPO_4^{2-}] + [PO_4^{3-}]$ (7)

C.B.: $[Na+] + [H^+] = 0.3 + [H^+] = [OH^-] + [H_2PO_4^-] + 2[HPO_4^{2-}] + 3[PO_4^{3-}]$ (8)

or $0.3 = [OH^-] - [H^+] + [H_2PO_4^-] + 2[HPO_4^{2-}] + 3[PO_4^{3-}]$ (8a)

Multiplying Equation (7) by 3 and combining the result with Equation (8a),
$3[H_3PO_4] + 3[H_2PO_4^-] + 3[HPO_4^{2-}] + 3[PO_4^{3-}] = [OH^-] - [H^+] + [H_2PO_4^-] + 2[HPO_4^{2-}] + 3[PO_4^{3-}]$

or $[H_2PO_4^-] = \dfrac{[OH^-] - [H^+] - [H_2PO_4^-] - 3[H_3PO_4]}{2}$

(f)   M.B.: $0.1 = [HSO_4^-] + [SO_4^{2-}]$ (because $[H_2SO_4] = 0$)   (9)
      C.B: $[H^+] = [OH^-] + [HSO_4^-] + 2[SO_4^{2-}]$ (10)

Combining Equations (9) and (10), we have
$[HSO_4^-] = 0.1 - [SO_4^{2-}] = 0.1 - ([H^+] - [OH^-] - [HSO_4^-])/2$
or $[HSO_4^-] = 0.2 - [H^+] + [OH^-]$

11. If S is the molar solubility of BaF_2, then
$S = [Ba^{2+}]$
$2S = [F^-] + [HF] + 2[HF_2^-]$ $\therefore$ $[F^-] + [HF] + 2[HF_2^-] = 2[Ba^{2+}]$

12. $2[Ba^{2+}] = 3([PO_4^{3-}] + [HPO_4^{2-}] + [H_2PO_4^-] + [H_3PO_4])$

13. Equilibria:
$HOAc = H^+ + OAc^-$
$H_2O = H^+ + OH^-$

Equilibrium expressions
$K_a = [H^+][OAc^-]/[HOAc] = 1.75 \times 10^{-5}$ (1)
$K_w = [H^+][OH^-] = 1.00 \times 10^{-14}$ (2)

Mass balance expressions
$C_{HOAc} = [HOAc] + [OAc^-] = 0.100\ M$ (3)
$[H^+] = [OAc^-] + [OH^-]$ (4)

Charge balance expression
$[H^+] = [OAc^-] + [OH^-]$ (5)

Number of expressions vs. number of unkowns:
There are four unkonwns ($[HOAc]$, $[OAc^-]$, $[H^+]$, $[OH^-]$) and four expressions (two equilibrium and two mass balance – the charge balance expression is the same as (4)).

Simplifying assumptions:
In acid solution, $[OH^-] \ll [H^+]$.

From (4): $[OAc^-] \approx [H^+]$
From (3): $[HOAc] = 0.100 - [OAc^-] = 0.100 - [H^+]$
From (1):
$[H^+]^2/(0.100 - [H^+]) = 1.75 \times 10^{-5}$
$[H^+] = 1.31 \times 10^{-3}$
pH = 2.88

14 (a) $\mu = ([Na^+](1)^2 + [Cl^-](1)^2)/2 = [(0.30)(1) + (0.30)(1)]/2 = 0.30$

(b) $\mu = ([Na^+](1)^2 [SO_4^{2-}](2)^2)/2 = [(0.60)(1) + (0.30)(4)]/2 = 0.90$

(c) $\mu = ([Na^+](1)^2 + [Cl^-](1)^2 + [K^+](1)^2 + [SO_4^{2-}](2)^2)/2$

$= [(0.30)(1) + (0.30)(1) + (0.40)(1) + (0.20)(4)]/2 = 0.90$

(d) $\mu = ([Al^{3+}](3)^2 + [SO_4^{2-}](2)^2 + [Na^+](1)^2/2$

$= [(0.40)(9) + (0.60 + 0.10)(4) + (0.20)(1)]/2 = 3.3$

15. (a) $\mu = ([Zn^{2+}](2)^2 + [SO_4^{2-}](2)^2)/2 = [(0.20(4) + (0.20)(4)]/2 = 0.80$

(b) $\mu = ([Mg^{2+}](2)^2 + [Cl^-](1)^2)/2 = [(0.40)(4) + (0.80)(1)]/2 = 1.20$

(c) $\mu = ([La^{3+}](3)^2 + [Cl^-](1)^2)/2 = [(0.50)(9) + (1.50)(1)]/2 = 3.0$

(d) $\mu = [K^+](1)^2 + [Cr_2O_7^{2-}](2)^2/2 = [(2.0)(1) + (1.0)(4)]/2 = 3.0$

(e) $\mu = ([Tl^{3+}](3)^2 + [Pb^{2+}](2)^2 + [NO_3^-](1)^2)/2$

$= [(1.0)(9) + (1.0)(4) + (5.0(1)]/2 = 9.0$

16. $\mu = [NaCl] = 1.00 \times 10^{-3}$. This is less than 0.01, so use Equation 6.20.

$-\log f_{Na^+} = -\log f_{Cl^-} = [(0.51)(1)^2(1.00 \times 10^{-3})^{1/2}]/[1 + (1.00 \times 10^{-3})^{1/2}] = 0.015_6$

$f_{Na^+} = f_{Cl^-} = 10^{-0.015}{}_6 = 10^{-1} \times 10^{.984}{}_4 = 0.96_5$

17. $\mu = [(0.0040)(1)^2 + (0.0050)(2)^2 + (0.0020)(3)^2]/2 = 0.021$

This is >0.01, so use Equation 6.19. From Recommended Reference 9 in Chapter 6,

$\alpha_{Na^+} = 4$, $\alpha_{Al^{3+}} = 9$, $\alpha_{SO_4^{2-}} = 4$

$-\log f_{Na^+} = [(0.51)(1)^2(0.021)^{1/2}]/[1 + (0.33)(4)(0.021)^{1/2}] = 0.062$

$f_{Na^+} = 0.867$

$-\log f_{SO_4^{2-}} = [(0.51)(2)^2(0.021)^{1/2}]/[1 + (0.33)(4)(0.021)^{1/2}] = 0.25$

$f_{SO_4^{2-}} = 0.56$

$-\log f_{Al^{3+}} = [(0.51)(3)^2(0.021)^{1/2}]/[1 + (0.33)(9)(0.021)^{1/2}] = 0.46$

$f_{Al^{3+}} = 0.35$

18. $\mu = [(0.0020)(1)^2 + (0.0020)(1)^2]/2 = 0.0020$

Since $\mu < 0.01$, use Equation 6.20:

$-\log f_{NO_3^-} = [(0.51)(1)^2(0.0020)^{1/2}]/[1 + (0.0020)^{1/2}] = 0.022$

$f_{NO_3^-} = 0.95$

$a_{NO_3^-} = ((0.0020)(0.95) = 0.0019 \ M$

19. $\mu = [(0.040)(1)^2 + (0.020)(2)^2]/2 = 0.060$

Since $\mu > 0.01$, use Equation 6.19. From Recommended Reference 10 in Chapter 6,

$\alpha_{CrO_4^{2-}} = 4$

$\log f_{CrO_4^{2-}} = [(0.51)(2)^2(0.060)^{1/2}]/[1 + (0.33)(4)(0.060)^{1/2}] = 0.37_8$

$f_{CrO_4^{2-}} = 0.42$

$a_{CrO_4^{2-}} = (0.020)(0.42) = 0.0084 \ M$

20. The solution is 2.5 molar, 2.5 moles per liter. If the density is 1.15, 1 liter is 1150 g.
    H_2SO_4 has a molecular weight of 98.1, 2.5 mol is 245.2 g. Amount of water = 1150 – 245.2
    = 904.8 g
    2.5 mol/0.9048 kg = 2.76 m (molal)
    At this concentration the second proton will not dissociate significantly, H_2SO_4 will only
    dissociate into H^+ and HSO_4^-. It will be treated as a 1:1, not a 2:1 electrolyte.
    Therefore $v = 2$ and $\mu = 2.5$, $\sqrt{\mu} = 1.58$
    The relative humidity tells us that a_w is 0.888
    Using 6.22. here Z_{H+} and Z_{HSO4-} are both 1. The ion size parameter for HSO_4^- is not available;
    we assume this is the same as for SO_4^{2-}, 4 A. The ion size parameter for H^+ is 9 A. The
    mean ion size parameter a_i is thus 6.5 A.

    $$-\log f\pm = \frac{0.51 Z_A Z_B \sqrt{\mu}}{1 + 0.33 a_i \sqrt{\mu}} + n/(\log a_w + \log(1 - 0.018(n-v)m)$$

    $\quad\quad\quad = (0.51*1.58)/(1 + 0.33*6.5*1.58) + (4/2)\log(0.888) + \log (1-0.018*(4-2)*2.76)$
    $\quad\quad\quad = 0.806/4.39 + 2*(-.052) +\log(1 - 0.0994)$
    $\quad\quad\quad = 0.183 - 0.104 -.045 = 0.034$
    $\quad\quad f_\pm = 10^{-0.034} = 0.925$

21. (a) $K_a^0 = (a_{H+}.a_{CN-})/a_{HCN} = ([H^+]f_{H+}.[CN^-]f_{CN-}/[HCN] = K_a\, f_{H+}f_{CN-}$

    (b) $K_b^0 = (a_{NH4+}.a_{OH-})/a_{NH3} = ([NH_4^+]f_{NH4+}.[OH^-]f_{OH-}/[NH_3] = K_b\, f_{NH4+}f_{OH-}$

22. (a) $HBenz = H^+ + Benz^-$
    From Appendix C (constants at $\mu = 0$)
    $K_a^0 = [H^+][Benz^-]/[Hbenz] = 6.3 \times 10^{-5}$
    $(x)(x)/(5.0 \times 10-3) = 6.3 \times 10^{-5}$ (Neglect x compared to C, which is $\approx 100 \times K_a^0$)
    $x = 5.6 \times 10^{-4}\, M = [H^+]$
    $pH = -\log (5.6 \times 10^{-4}) = 3.25$

    (b) $\mu = [(0.100)(1)^2 + (0.050)(2)^2]/2 = 0.15$

    From Recommended Reference 9 in Chapter 6, $\alpha_{H+} = 9$.
    $-\log f_{H+} = [(0.51)(1)^2(0.15)^{1/2}]/[1 + (0.33)(9)(0.15)^{1/2}] = 0.092$
    $f_{H+} = 0.81$

    From Recommended Reference 9 in Chapter 6, $\alpha_{Benz-} = 6$.
    $-\log f_{Benz-} = [(0.51)(1)^2(0.15)^{1/2}]/[1 + (0.33)(6)(0.15)^{1/2}] = 0.11_1$
    $f_{Benz-} = 0.77$
    At $\mu = 0.15$,
    $K_a = K_{a0}(f_{HA}/f_{H+}f_{A-}) = K_{a0}(1/f_{H+}f_{A-})$
    $K_a = (6.3 \times 10^{-5})/(0.81)(0.77) = 1.0_1 \times 10^{-4}$
    $(x)(x)/(5.0 \times 10^{-3} - x) = 1.0_1 \times 10^{-4}$
    Solving quadratic formula,
    $x = 6.6 \times 10^{-4} M = [H^+]$
    $pH = -\log(6.6 \times 10^{-4}) = 3.18$ [paH = $-\log(6.6 \times 10^{-4} \times 0.81) = 3.28$]

23. See the text website for the spreadsheet. The formula for f_i is:

$f_i = 10^{\wedge} -(0.51 * Z_{i\ cell}{}^{\wedge}2 * \mu_{cell}{}^{\wedge}0.5)/(1 + \mu_{cell}{}^{\wedge}0.5)$

24. See the text website for the spreadsheet. The spreadsheet calculated values are 0.919 for K^+ (vs. 0.918) and 0.713 for $SO_4{}^{2-}$ (vs. 0.713). The slight difference is due to rounding in the manual calculation.

25. See the text website for the spreadsheets (a and b). The spreadsheet values are 0.794 for K^+ (vs. 0.792) and 0.419 for $SO_4{}^{2-}$ (vs. 0.419).

26. See the text website for the Goal Seek setup and solution. The calculated answer is 6.18 x 10^{-4} M for x.

27. See the text website for the spreadsheet calculations, Problems 16, 17, 18, and 19.

CHAPTER 7   ACID−BASE EQUILIBRIA

1. A strong electrolyte is completely ionized in solution, while a weak electrolyte is only partially ionized in solution. A slightly soluble salt is generally a strong electrolyte, because it is completely ionized in solution.

2. The Brønsted acid-base theory assumes that an acid is a proton donor, and a base is a proton acceptor. In the Lewis theory, an acid is an electron acceptor, while a base is an electron donor.

3. A conjugate acid is the protonated form of a Brønsted base, and a conjugate base is the ionized form of a Brønsted acid: conjugate acid = H^+ + conjugate base.

4. $C_6H_5NH_2 + HOAc \rightarrow C_6H_5NH_3^+ + OAc^-$
    conjugate acid

    $C_6H_5NH_2 + NH_2CH_2CH_2NH_2 \rightarrow C_6H_5O^- + NH_2CH_2CH_2NH_3^+$
    conjugate base

5. Good buffers are designed for use in biological systems, in the biologically important pH range of 6-8. They possess properties that will not interfere with biological processes. Tris buffers are examples.

6. (a) pH = -log 2.0×10^{-2} = 2 – 0.30 = 1.70
    pOH = 14.00 – 1.70 = 12.30

    (b) pH = -log 1.3×10^{-4} = 4 – 0.11 = 3.89
    pOH = 14.00 – 3.89 = 10.11

    (c) pH = -log 1.2 = -0.08
    pOH = 14.00 –(-0.08) = 14.08

    (d) The concentration of HCl is about 100 times less than the concentration of H^+ for the ionization of water. Therefore, the former can be neglected and pH = pOH = 7.00. (pH ≠ $-\log 1.2 \times 10^{-9}$ = 8.92, which is alkaline!)

    (e) The contribution from water is appreciable. Therefore, use K_w to calculate its contribution.

    $$H_2O \quad = \quad H^+ \quad + \quad OH^-$$
    $$(2.4 \times 10^{-7} + x) \qquad x$$
    $$(2.4 \times 10^{-7} + x)(x) = 1.0 \times 10^{-14}$$

    From the quadratic equation, x = $3._5 \times 10^{-8}$ M
    $[H^+]$ = 2.4×10^{-7} + 0.4×10^{-7} = 2.8×10^{-7} M
    pH = -log 2.8×10^{-7} = 7 – 0.45 = 6.55 (slightly acid)
    pOH = 14.00 – 6.55 = 7.45

7. (a) pOH = -log 5.0 x 10^{-2} = 2 – 0.70 = 1.30
       pH – 14.00 – 1.30 = 12.70

   (b) pOH = -log 2.8 x 10^{-1} = 1 – 0.45 = 0.55
       pH = 14.00 – 0.55 = 13.45

   (c) pOH = -log 2.4 = -0.38
       pH = 14.00 – (-0.38) = 14.38

   (d) Must calculate contribution from H_2O
       H_2O = H^+ + OH^-
                       x   3.0 x 10^{-7} + x
       (x)(3.0 x 10^{-7} + x) = 1.0 x 10^{-14}

       From the quadratic equation: x = 1 x 10^{-8} M
       $[OH^-]$ = 3.0 x 10^{-7} x 0.2 x 10^{-7} = 3.2 x 10^{-7} M
       pOH = -log 3.2 x 10^{-7} = 7 –0.51 = 6.49
       pH = 14.00 – 6.49 = 7.51

   (e) pOH = -log 3.7 x 10^{-3} = 3 –0.57 = 2.43
       pH = 14.00 – 2.43 = 11.57

8. (a) $[OH^-]$ = (1.0 x 10^{-14})/(2.6 x 10^{-5}) = 3.8 x 10^{-10} M

   (b) $[OH^-]$ = (1.0 x 10^{-14})/(0.20) = 5.0 x 10^{-14} M

   (c) $[H^+]$ = 1.0 x 10^{-7} M ($HClO_4$ is negligible)
       $[OH^-]$ = 1.0 x 10^{-7} M

   (d) $[OH^-]$ = (1.0 x 10^{-14})/(1.9) = 5.3 x 10^{-15} M

9. (a) $[H^+]$ = $10^{-3.47}$ = 10^{-4} x $10^{.53}$ = 3.4 x 10^{-4} M

   (b) $[H^+]$ = $10^{-0.20}$ = 10^{-1} x $10^{.80}$ = 6.3 x 10^{-1} M (0.63 M)

   (c) $[H^+]$ = $10^{-8.60}$ = 10^{-9} x $10^{.40}$ = 2.5 x 10^{-9} M

   (d) $[H^+]$ = $10^{-(-.60)}$ = $10^{.60}$ = 4.0 M

   (e) $[H^+]$ = $10^{-14.35}$ = 10^{-15} x $10^{.65}$ = 4.5 x 10^{-15} M

   (f) $[H^+]$ = $10^{1.25}$ = 10^1 x $10^{.25}$ = 1.8 x 10^1 M (18 M)

10. Assume 1.0 mL of each is mixed.
    Excess NaOH = $(0.30\ M \times 1.0\ mL - 0.10\ M \times 1.0\ mL \times 2)/2\ mL$
    = 0.10 mmol/2 mL = 0.050 M
    pOH = -log 5.0 x 10^{-2} = 2 – 0.70 = 1.30
    pH = 14.00 – 1.30 = 12.70

11. Assume 1.0 mL volumes.
    $[H^+]$ of acid solution = 1.0 x 10^{-3} M
    $[H^+]$ of base solution = 1.0 x 10^{-12} M
    $[OH^-]$ = (1.0 x 10^{-14})/(1.0 x 10^{-12}) = 1.0 x 10^{-2} M
    Excess base = (1.0 x 10^{-2} M x 1.0 mL - 1.0 x 10^{-3} M x 1.0 mL)/2 mL
    = 9 x 10^{-3} mmol/2 mL = 4.5 x 10^{-3} M OH^-
    $\therefore$ pOH = -log 4.5 x 10^{-3} = 3 – 0.65 = 2.35
    pH = 14.00 – 2.35 = 11.65

12. PFP
    In this case the condition for neutrality is that
    $n_a = n_b$
    where n_a is mmoles of strong acid; n_b=mmoles of strong base.

    Neglecting ionic strength effects, a pH= 2.00 strong acid solution has 0.010 M H^+ while a
    pH=11.00 solution has 0.0010 M OH^-. Thus,
    0.010 V_a= 0.0010 V_b
    or
    V_a= 0.10 V_b

    In other words, since the strong acid solution is ten times more concentrated than the strong
    base solution, you need 10 times more volume of strong base to compensate for the
    difference in concentration.

13. $[H^+][OH^-]$ = 5.5 x 10^{-14}
    $[H^+]^2$ = 5.5 x 10^{-14}
    $[H^+]$ = 2.3 x 10^{-7} M
    pH = -log 2.3 x 10^{-7} = 7 – 0.36 = 6.64

14. pH + pOH = 13.60
    pOH = 13.60 –7.40 = 6.20

15. $[H^+]$ = $[OAc^-]$ = $10^{-3.26}$ = 10.74 x 10^{-4} = 5.5 x 10^{-4} M
    (5.5 x 10^{-4})^2/[HOAc] = 1.75 x 10^{-5}
    [HOAc] = 1.7_3 x 10^{-2} M (neglecting $[H^+]$ in the denominator).
    % ionized = [(5.5 x 10^{-4})/(1.7_3 x 10^{-2})] x 100% = 3.2%

16. PFP

    (a) K_b= K_w/K_a = 5.71×10^{-10}

(b) $CH_3COO^- + H_2O \leftrightarrow CH_3COOH + OH^-$

$$K_b = \frac{[CH_3COOH][OH^-]}{[CH_3COO^-]} \; , \; [OH^-] = \sqrt{K_b \times 0.1} = 7.56 \times 10^{-6}$$

$$pOH = -\log[OH^-] = 5.12, \; pH = 14 - pOH = 8.88$$

17. $pOH = 14.00 - 8.42 = 5.58$
    $[OH^-] = [RNH_3^+] = 10^{-5.58} = 10^{.42} \times 10^{-6} = 2.6 \times 10^{-6} \, M$
    $K_b = (2.6 \times 10^{-6})^2/(0.20) = 3.4 \times 10^{-11}$
    $pK_b = -\log 3.4 \times 10^{-11} = 10.47$

18. Let x = concentration of acid
    $(0.035x)^2/(x) = 6.7 \times 10^{-4}$
    $x = 0.55 \, M$
    $100 \, g/L = 0.55 \, mol/L$
    $\therefore$ f.w. = 100 g/0.55 mol = 18_2 g/mol

19. $[H^+][Prop^-]/[Hprop] = 1.3 \times 10^{-5}$
    $(x)(x)/(0.25) = 1.3 \times 10^{-5}$
    $x = 1.3 \times 10^{-3}$
    $pH = -\log 1.3 \times 10^{-3} = 3 - 0.26 = 2.74$

20. $RNH_2 + H_2O = RNH_3^+ + OH^-$
    $K_b = [RNH_3^+][OH^-]/[RNH_2]$
    $4.0 \times 10^{-10} = (x)(x)/(0.10)$
    $x = 6.3 \times 10^{-6}$
    $pOH = -\log 6.3 \times 10^{-6} = 6 - 0.80 = 5.20$
    $pH = 14.00 - 5.20 = 8.80$

21. $[H^+][IO_3^-]/[HIO_3] = 2 \times 10^{-1}$
    $HIO_3 \; = \; H^+ \; + \; IO_3^-$
    $0.1 - x \quad\quad x \quad\quad x$
    $(x)(x)/(0.1 - x) = 2 \times 10^{-1}$

    Since K_a is large, the quadratic equation must be solved:
    $x^2 = 0.20 - 0.2x$
    $x^2 + 0.2x - 0.02 = 0$
    $x = [-0.2 \pm \sqrt{(0.2)^2 - (4)(1)(0.02)}]/(2 x 1)$
    $x = 0.07 \, M = [H^+]$
    $pH = -\log 0.07 = 2 - 0.8 = 1.2$

22. $H_2SO_4 \rightarrow H^+ + HSO_4^-$
    $H_2SO_4 \rightarrow H^+ + HSO_4^-$

    The total $[H^+] = [HSO_4^-] + [SO_4^{2-}]$. Let x = $[SO_4^{2-}]$. Then
    $[HSO_4^-] = 0.0100 - x$

$[H^+] = 0.0100 + x$

$K_{a2} = [H^+][SO_4^{2-}]/[HSO_4^-] = 1.2 \times 10^{-2}$

K_{a2} is not very small, so x is probably appreciable and can't be neglected.

$\therefore \ (0.0100) + x)(x)/(0.0100 - x) = 1.2 \times 10^{-2}$

The quadratic formula must be used to solve for x:

$x^2 + 0.022x - 1.2 \times 10^{-4} = 0$

$x = [-0.022 \pm \sqrt{(0.022)^2 - (4)(-1.2x10^{-4})}]/2 = 0.004_5 \ M$

$[H^+] = 0.0100 + 0.004_5 = 0.014_5 \ M$

23. From the Appendix, $K_a = 0.129$

$$HT \quad = \quad H^+ \quad + \quad T^-$$
$$0.100 - x \quad \quad x \quad \quad x$$

The initial concentration is much less than $100K_a$, so the quadratic equation must be solved.

$[H^+][T^-]/[HT] = 0.129$

$(x)(x)/0.100 - x) = 0.129$

$x^2 + 0.129x - 0.0129 = 0$

$x = [-0.129 \pm \sqrt{(0.129)^2 - (4)(-0.0129)}]/2 = 0.0661 \ M$

24. $RNH_2 + H_2O = RNH_3^+ + OH^-$
$$\quad 0.20 - x \quad \quad \quad \quad x \quad \quad \quad x$$
$K_b = 10^{-4.20} = 10^{-5} \times 10^{.80} = 6.3 \times 10^{-5}$

$[RNH_3^+][OH^-]/[RNH_2] = 6.3 \times 10^{-5}$

$(x)(x)/0.20 = 6.3 \times 10^{-5}$

$x = 3.5 \times 10^{-3} \ M = [OH^-]$

$pOH = -\log 3.5 \times 10^{-3} = 3 - 0.54 = 2.46$

$pH = 14.00 - 2.46 = 11.54$

25. $HOAc \quad = \quad H^+ \quad + \quad OAc^-$
$$\quad c - 0.030c \quad \quad 0.030c \quad \quad 0.030c$$
$$\approx c$$

$(0.030c)(0.030c)/c = 1.75 \times 10^{-5}$

$9.0 \times 10^{-4} c^2 = 1.75 \times 10^{-5} c$

$c = 0.019 \ M$

26. $HA = H^+ + A^-$
$$\quad c - x \quad x \quad x$$
$$\approx c$$
(since $c > 100 \ K_a$)

Assume 1% ionized at 0.100 M. Then $x = 0.00100$

$(0.00100)^2/(0.100) = K_a = 1.00 \times 10^{-5}$

For 2% ionized (doubled), x = 0.0200c

$(0.0200c)^2/c = 1.00 \times 10^{-5}$

$4.00 \times 10^{-4} c^2 = 1.00 \times 10^{-5} c$

c = 0.025 M

Therefore, must dilute HA 4-fold.

27. $H_3BO_3 + NaOH \rightarrow NaBO_2 + 2H_2O$

mmol H_3BO_3 = 20 mL x 0.25 M = 5.0 mmol

mmol NaOH = 25 mL x 0.20 M = 5.0 mmol

These are stoichiometrically equal, so 5.0 mmol BO_2^- is formed in 45 mL.

$[BO_2^-]$ = 6.0 mmol/45 mL = 0.11_1 mmol/mL

$\qquad BO_2^- \;\;+\;\; 2H_2O \;\;=\;\; H_3BO_3 \;\;+\;\; OH^-$

$0.11_1 - x \qquad\qquad\qquad\quad x \qquad\quad x$

$[H_3BO_3][\,OH^-]/[\,BO_2^-] = K_b = K_w/K_a = (1.0 \times 10^{-14})/(6.4 \times 10^{-10}) = 1.6 \times 10^{-5}$

c > 100 K_b. ∴ c ≈ ≈0.11_1 M

$(x)(x)/(0.11_1) = 1.6 \times 10^{-5}$

  x = $1.3_3 \times 10^{-3}$ M = $[OH^-]$

Or, from Equation 7.32,

$[OH^-] = \sqrt{K_b.C_A^-} = \sqrt{(1.6 x 10^{-5})(0.11_1)} = 1.3_3 \times 10^{-3}$ M

pOH = 2.88; pH = 14.00 − 2.88 = 11.12

28. $\quad CN^- \;\;+\;\; H_2O \;\;=\;\; HCN \;\;+\;\; OH^-$

  $0.010 - x \qquad\qquad\quad x \qquad\quad x$

$[HCN][OH^-]/[CN^-] = K_b = K_w/K_a = (1.0 \times 10^{-14})/(7.2 \times 10^{-10}) = 1.4 \times 10^{-5}$

C_{CN^-} > 100 K_b. ∴ neglect x compared to 0.010 M and use Equation 7.32.

$[OH^-] = \sqrt{K_b.C_A^-} = \sqrt{(1.4 x 10^{-5})(0.010)} = 3.7 \times 10^{-4}$ M

pOH = 3.43; pH = 14.00 − 3.43 = 10.57

29. K_a benzoic acid = 6.3×10^{-5}

$C_6H_6C00^- + H^2O = C_6H_6C00H + OH^-$

From Equation 7.32 (since C_A^- > 200 K_b),

$[OH^-] = \sqrt{K_w / K_a[C_6H_5COO^-]}$

$= \sqrt{[(1.0 x 10^{-14})/(6.3 x 10^{-5})]x0.050} = 2.8 \times 10^{-6}$ M

pOH = -log 2.8 x 10^{-6} = 5.55

pH = 14.00 − 5.55 = 8.45

30. K_b pyridine = 1.7×10^{-9}

$C_5H_5NH^+ + H_2O = C_6H_5NHOH + H^+$; C_{BH^+} > 100 K_a, so use Equation 7.39,

$[H^+] = \sqrt{K_w / K_b[C_6H_5NH^+]} = \sqrt{[(1.0 x 10^{-14})/(1.7 x 10^{-9})]x0.25} = 1.2 \times 10^{-3}$ M

pH = -log 1.2 x 10^{-3} = 2.92

31. mmol H$^+$ = 0.25 M x 12.0 mL x 2 = 6.0 mmol
    mmol NH$_3$ = 1.0 M x 6.0 mL = 6.0 mmol
    $\therefore$ 6.0 mmol of NH$_4^+$ are formed in a volume of 18.0 mL.
    [NH$_4^+$] = 6.0 mmol/18 mL = 0.33 M
    [H$^+$] = $\sqrt{[(1.0x10^{-14})/(1.75x10^{-5})]x0.33}$ = 1.4 x 10^{-5} M
    pH = 5 – log 1.5 = 5 – 0.15 = 4.85

32. mmol HOAc = 0.10 M x 20 mL = 2.0 mmol
    mmol NaOH = 0.10 M x 20 mL = 2.0 mmol
    $\therefore$ 2.0 mmol of NaOAc are formed in a volume of 40 mL.
    [OAc$^-$] = 2.0 mmol/40 mL = 0.050 M
    [OH$^-$] = $\sqrt{[(1.0x10^{-14})/(1.75x10^{-5})]x0.050}$ = 5.3 x 10^{-6}
    pOH = 14.00 – 5.28 = 8.72

33. HONH$_2$ + HCl = HONH$_3^+$ + Cl$^-$
    We form 0.10 mol HONH$_3^+$/0.50 L = 0.20 M
    HONH$_3^+$ + H$_2$O = HONH$_3$OH + H$^+$
    [HONH$_3$OH][H$^+$]/[HONH$_3^+$] = K$_a$ = K$_w$/K$_b$ = 1.0 x 10^{-14}/9.1 x 10^{-9} = 1.1 x 10^{-6}
    C$_{BH+}$ > 100 K$_a$, therefore,
    $\quad$ [H+] = $\sqrt{K_a.C_{BH}^+}$ = $\sqrt{(1.1x10^{-6})(0.20)}$ = 4.7 x 10^{-4} M
    pH = 3.33

34. C$_6$C$_5$(OH)COO$^-$ + H$_2$O = C$_6$C$_5$(OH)COOH + OH$^-$
    $\quad$ c – x $\qquad\qquad\qquad$ x $\qquad\qquad$ x
    [C$_6$C$_5$(OH)COOH][OH$^-$]/[C$_6$C$_5$(OH)COO$^-$] = K$_b$ = K$_w$/K$_a$= 1.0 x 10^{-14}/9.1 x 10^{-3}
    = 1.0 x 10^{-11}
    $\quad$ C$_{A-}$ > 100 K$_b$, therefore,
    [OH$^-$] = $\sqrt{(1.0x10^{-11})(0.0010)}$ = 1.0 x 10^{-7} M
    pOH = pH = 7.00

35. $\quad$ CN$^-$ $\quad$ + $\quad$ H$_2$O = HCN $\quad$ + $\quad$ OH$^-$
    $\quad$ 1.0 x 10^{-4} – x $\qquad\qquad\qquad$ x $\qquad\qquad$ x

    From Problem 26, K$_b$ = 1.4 x 10^{-5}. C$_{A-}$ < 200 K$_b$, so we can't neglect x.
    Therefore, solve the quadratic formula.
    This gives for (x)(x)/(1.0 x 10^{-4} – x) = 1.4 x 10^{-5}
    x = [OH$^-$] = 3.1 x 10^{-5} M (the hydrolysis is 31% complete)
    pOH = 4.51; pH = 9.49

36. $\quad$ H$_2$P $\quad$ = $\quad$ H$^+$ $\quad$ HP$^-$
    $\quad$ 0.0100 – x $\qquad$ x $\qquad$ x

    Assume second ionization is small.
    $\quad$ [H$^+$][HP$^-$]/[H$_2$P] = K$_{a1}$

$(x)(x)/(0.0100 - x) = 1.2 \times 10^{-3}$
$c < 200 K_a$. Therefore must solve by quadratic equation.
$x = [-1.2x10^{-3} \pm \sqrt{(1.2x10^{-3})^2 - (4)(-1.2x10^{-5})}]/2 = 2.9 \times 10^{-3} M$
$pH = -\log 2.9 \times 10^{-3} = 2.54$

37. P^{2-} + H_2O = HP^- + OH^-
     $0.0100 - x$ x x

Assume further hydrolysis of HP^- is negligible.
$[HP^-][[OH^-]/[P^{2-}] = K_b = (K_w/K_{a2}) = (1.0 \times 10^{-14})(3.9 \times 10^{-6}) = 2.6 \times 10^{-9}$
K_b is small, so x can be neglected compared to 0.0100 M, and the quadratic equation need not be used.

$[OH^-] = \sqrt{(K_w/K_{a2}) x[P^{2-}]} = \sqrt{2.6x10^{-9}x1.00x10^{-2}} = 5.1 \times 10^{-6} M$
$pOH = 5.29; pH = 14.00 - 5.29 = 8.71$

38. This is an amphoteric salt:
$HP^- = H^+ + P^{2-}$
$HP^- + H_2O = H_2P + OH^-$

$[H^+] = \sqrt{K_{a1}K_{a2}} = \sqrt{1.2x10^{-3}x3.9x10^{-6}} = 6.9 \times 10^{-5} M$
$pH = -\log 6.9 \times 10^{-5} = 5 - 0.84 = 4.16$

39. S^{2-} hydrolyzes:
        S^{2-} + H_2O = HS^- + OH^-
     $0.600 - x$ x x
$K_b = (K_w/K_{a2}) = (1.0 \times 10^{-14})/(1.2 \times 10^{-15}) = 8.3$
The equilibrium therefore lies significantly to the right and the quadratic equation must be used to solve the for the degree of ionization:
$[HS^-][OH^-]/[S^{2-}] = (x)(x)/(0.600 - x) = 8.3$
$x^2 + 8.3x - 4.9_8 = 0$
$x = [-8.3 \pm \sqrt{(8.3)^2 - (4)(-4.9_8)}]/2 = 0.55 M$ (92% hydrolyzed! Strong base)
$pOH = -\log 0.55 = 0.26; pH = 14.00 - 0.26 = 13.74$

40. PO_4^{3-} hydrolyzes. From Equation 7.87, $K_b = (K_w/K_{a3}) = [HPO_4^{2-}][OH^-]/[PO_4^{3-}]$
    $= (1.0 \times 10^{-14})/(4.8 \times 10^{-13}) = 0.021$

Since the equilibrium constant is not very small, the quadratic equation must be used to solve for the degree of ionization. If
$x = [OH^-] = [HPO_4^{2-}]$, and $[PO_4^{3-}] = 0.500 - x$, then:
$(x^2)/(0.500 - x) = 0.021$
$x^2 + 0.021x - 0.010_5 = 0$

$x = [-0.021 \pm \sqrt{(0.021)^2 - (4)(-0.010_5)}]/2 = 0.092 M$
$pOH = -\log 0.092 = 1.04; pH = 14.00 - 1.04 = 12.96$

41. HCO_3^- is amphoteric. For H_2CO_3, $K_{a1} = 4.3 \times 10^{-7}$ and $K_{a2} = 4.8 \times 10^{-11}$. Since the difference between these is large, we can write for HCO_3^-:

$[H^+] = \sqrt{K_{a1}K_{a2}} = \sqrt{4.3x10^{-7} x4.8x10^{-11}} = 4.5 \times 10^{-9} \ M$
$pH = -\log 4.5 \times 10^{-9} = 8.34$

42. HS^- is amphoteric.

$[H^+] = \sqrt{K_{a1}K_{a2}} = \sqrt{9.1x10^{-8} x1.2x10^{-15}} = 1.04 \times 10^{-11} \ M$
$pH = -\log 1.04 \times 10^{-11} = 10.98$

43. $HY^{3-} = H^+ + Y^{4-}$　　　　　$K_{a4} = 5.5 \times 10^{-11}$
$HY^{3-} + H_2O = H_2Y^{2-} + OH^-$　　$K_b = K_w/K_{a3} = 1.0 \times 10^{-14}/6.9 \times 10^{-7}$
$= 1.4 \times 10^{-8}$ (use K_a of the conjugate acid, H_2Y^{2-})

The difference between K_{a3} and K_{a4} is large, so
$[H^+] = \sqrt{K_{a3}K_{a4}} = \sqrt{6.9x10^{-7} x5.5x10^{-11}} = 6.2 \times 10^{-9} \ M$
$pH = 8.21$

44. PFP
(a) (i) A^{2-} ; (ii) HA^- ; (iii) H_2A (b) (i) HA^{2-} (ii) A^{3-} (iii) H_2A^-

45. $pH = pK_a + \log ([COO^-]/[HCOOH]);$ $pKa = 3.75$
$pH = 3.75 + \log (0.10/0.050) = 3.75 + 0.30 = 4.05$

46. $mmol \ NH_3 = 0.10 \times 5.0 = 0.50 \ mmol$
$mmol \ HCl = 0.020 \times 10l0 = 0.20 \ mmol$
$mmol \ excess \ NH_3 = 0.50 - 0.20 = 0.30 \ mmol$
$pH = pK_a + \log [proton \ acceptor]/[proton \ donor] = (pK_w - pK_b) + \log [NH_3]/[NH_4^+]$
$= (14.00 - 4.76) + \log (0.30)/(0.20) = 9.24 + 0.18 = 9.42$
Or, from Equation 7.58,
$pOH = 4.76 + \log [NH_4^+]/[NH_3] = 4.76 + \log (0.20/0.30) = 4.58$
$pH = 14.00 - 4.58 = 9.42$

47. $5.00 = 4.76 + \log (0.100/[HOAc])$
$[HOAc] = 0.058 \ M$
$mmol \ HOAc = 0.058 \ M \times 100 \ mL = 5.8 \ mmol$
$mmol \ NaOAc = 0.100 \ M \times 100 \ mL = 10.0 \ mmol$
$mmol \ NaOH \ added = 0.10 \ M \times 10 \ mL = 1.0 \ mmol$

After adding NaOH:
$mmol \ NaOAc = 10.0 + 1.0 = 11.0$
$mmol \ HOAc = 5.8 - 1.0 = 4.8$
$pH = 4.76 + \log (11.0)/4.8 = 5.12$
The pH increased by $5.12 - 5.00 = 0.12$

48. mmol HOAc = 50 mL x 0.10 M = 5.0 mmol
    mmol NaOH = 20 mL x 0.10 M = 2.0 mmol = mmol HOAc converted to OAc$^-$
    mmol HOAc left = 30 mmol
    pH = pK_a + log ([OAc$^-$]/[HOAc] = 4.76 + log (2.0/3.0) = 4.58

49. mmol NH$_3$ = 50 mL x .10 M = 5.0 mmol
    mmol H$_2$SO$_4$ = 25 mL x 0.050 M = 1.2$_5$ mmol x 2 = 2.5 mmol H$^+$ = mmol NH$_3$
    converted to NH$_4^+$
    mmol NH$_3$ left = 2.5 mmol
    pK_a = pK_w − pK_b = 14.00 − 4.76 = 9.24
    pH = pK_a + log [NH$_3$][NH$_4^+$] = 9.24 + log (2.5/2.5) = 9.24

50. From the pH of the stomach contents and the pK_a of aspirin, we can calculate the mole ratio
    of the ionized to un-ionized forms:
    pH = pK_a + log (mmol A$^-$/mmol HA)
    2.95 = 3.50 + log (mmol A$^-$/mmol HA)
    (mmol A$^-$/mmol HA) = 0.28; mmol A$^-$ = mmol HA x 0.28
    The total mmol of each form is 650 mg/180 (mg/mmol) = 3.6$_1$ mmol
    mmol HA + mmol A$^-$ = 3.6$_1$
    mmol HA + 0.28 x mmol HA = 3.6$_1$
    1.28 x mmol HA = 3.6$_1$
    mmol HA = 2.8$_2$
    mg HA = 2.8$_2$ mmol x 180 mg/mmol = 50$_8$ mg HA available for absorption (0.50$_8$ g)

51. pK_a = pK_w - pK_b
    pH = pK_a + log ([HOCH$_2$)$_3$NH$_2$]/[HOCH$_2$)$_3$NH$_3^+$])
    7.40 = 8.08 + log ([HOCH$_2$)$_3$NH$_2$]/[HOCH$_2$)$_3$NH$_3^+$])
    [HOCH$_2$)$_3$NH$_2$]/[HOCH$_2$)$_3$NH$_3^+$] = 0.21 = [THAM]/[THAMH$^+$]
    mmol HCl = 0.50 M x 100 mL = 50 mmol
    ∴ must add enough THAM to form 50 mmol THAMH$^+$, plus enough excess to satisfy the
    above ratio.
    Total THAM = 50 mmol + x mmol
    (x mmol/50 mmol) = 0.21
    x = 10.$_5$ mmol THAM
    Total THAM = 50 + 10 = 60 mmol
    mg THAM = 60 mmol x 121.1 mg/mmol = 7,300 mg = 7.3 g

52. In this case, K_a of the acid is large, and the presence of the added salt, T$^-$, will not suppress
    the ionization sufficiently to make x negligible.

    $$HT \quad = \quad H^+ \quad + \quad T^-$$
    $$0.100 - x \qquad x \qquad\quad x$$
    (x)(0.100 + x)/(0.100 − x) = 0.129
    x^2 + 0.229x = 0.0129 = 0
    $$x = [-0.229 \pm \sqrt{(0.229)^2 - (4)(-0.0129)}] / 2 = 0.46_7 \, M$$
    This compares with 0.129 M ([H$^+$] = K_a) had x been assumed negligible.

53. PFP

For C_6H_5COOH, $pK_a = 4.20$

$pH = pK_a + \log([C_6H_5COO^-]/[C_6H_5COOH])$

(a) $3.00 = 4.20 + \log([C_6H_5COO^-]/[C_6H_5COOH])$

$[C_6H_5COO^-]/[C_6H_5COOH] = 0.063$

Therefore, $[C_6H_5COOH]/[C_6H_5COO^-] = 16$

(b) $5.00 = 4.20 + \log([C_6H_5COO^-]/[C_6H_5COOH])$

$[C_6H_5COO^-]/[C_6H_5COOH] = 6.3$

Therefore, $[C_6H_5COOH]/[C_6H_5COO^-] = 0.16$

54. $pH = pK_{a1} + \log[(HP^-]/[H_2P])$

$K_{a1} = 1.2 \times 10^{-3}$; $pK_{a1} = 2.92$

$pH = 2.92 + \log(0.10/0.20) = 2.92 - 0.30 = 2.62$

55. $pH = pK_{a2} + \log([P^{2-}]/[HP^-])$

Since $[P^{2-}] = [HP^-]$,

$pH = pK_{a2} = -\log 3.9 \times 10^{-6} = 5.41$

56. $pH = pK_2 + \log([HPO_4^{2-}]/[H_2PO_4^-])$

$7.45 = 7.12 + \log([HPO_4^{2-}]/[H_2PO_4^-])$

$[HPO_4^{2-}]/[H_2PO_4^-]) = [H_2PO_4^-] = 3.0 \times 10^{-3}$

Solution of the two simultaneous equations gives:

$[HPO_4^{2-}] = 2.0 \times 10^{-3}$ M

$[H_2PO_4^-] = 9.6 \times 10^{-4}$ M

57. PFP

The formula weight of Na_2HPO_4 is 141.96 g/mol, and the formula weight of $NaH_2PO_4 \cdot H_2O$ is 137.99 g/mol.

For H_3PO_4, $pK_{a2} = 7.12$

$pH = pK_{a2} + \log([HPO_4^{2-}]/[H_2PO_4^-]) = 7.12 + \log[(0.6529 \text{ g} \div (141.96 \text{ g/mol} \times 0.100L)) / (0.2477 \text{ g} \div (137.99 \text{ g/mol} \times 0.100 \text{ L}))] = 7.53$

58. $\beta = 2.303 \, C_{HA}C_{A^-}/(C_{HA} + C_{A^-})$

$= 2.303 \times 0.10 \times 0.070/(0.10 + 0.070)$

$= 0.095$ mol/L per pH (or 10.5 pH per mol/L acid or base)

The volume change can be neglected when adding 10 mL acid or base.

The HCl and NaOH are diluted 1000-fold, to 0.0010 M.

For 10 mL 0.10 M HCl:

$dpH = -dC_A/\beta = -0.0010 \, M/0.095 \, M$ per pH $= -0.011$ pH $= \triangle$pH

For 10 mL 0.10 M NaOH,

$dpH = dC_B/\beta = 0.0010 \, M/0.095 \, M$ per pH $= +0.011$ pH $= \triangle$pH

59. At pH 4.76 (= pK_a), $[HOAc] = [OAc^-] = x$
    $\beta = 2.303\, C_{HA}C_{A^-}/(C_{HA} + C_{A^-})$
    1.0 M per pH = $2.303x^2/2x$
    $x = 0.868\ M = [HOAc] = [OAc^-]$

60. Let $x = M\ Na_2HPO_4$, $y = M\ KH_2PO_4$

    We have two unknowns and need two equations.
    $\mu = 1/2 \sum C_i Z_i^2$
    $0.20 = \frac{1}{2}\,([Na^+](1)^2 + [HPO_4^{2-}](2)^2 + [K^+](1)^2 + [H_2PO_4^-](1)^2)$
    $0.20 = \frac{1}{2}\,[2x(1)^2 + x(2)^2 + y(1)^2 + y(1)^2]$
    $0.20 = 3x + y$
    $pH = pK_2 + \log\,([HPO_4^{2-}]/[H_2PO_4^-])$
    $7.40 = 7.12 + \log\,(x/y)$

    Solution of the two simultaneous equations gives:

    $x = 0.057\ M\ Na_2HPO_4$
    $y = 0.030\ M\ KH_2PO_4$

    mg Na_2HPO_4 = 0.057 M x 200 mL x 142 mg/mmol = 1,620 mg = 1.6_2 g
    mg KH_2PO_4 = 0.030 M x 200 mL x 136 mg/mmol = 820 mg = 0.82 g

61. Let $x = M\ KH_2PO_4$, $y = M\ H_3PO_4$
    H_3PO_4 is non-ionic and does not contribute to the ionic strength.
    $\mu = 1/2 \sum C_i Z_i^2 = \frac{1}{2}\,([K+](1)^2 + [H_2PO_4^-](1)^2)$
    $\therefore\ 0.20 = \frac{1}{2}\,[x(1)^2 + x(1)^2] = x$

    $KH_2PO_4 = 0.20\ M$
    mg KH_2PO_4 = 0.20 M x 100 mL x 136 mg/mmol = 5,400 mg = 5.4 g
    $pH = pK_1 + \log\,([KH_2PO_4]/[H_3PO_4])$
    $3.00 = 1.96 + \log\,(0.20/y)$
    $y = 0.018\ M = 0.018\ M\ H_3PO_4$
    M conc. H_3PO_4 = (0.85 g/g soln x 1.69 g soln/mL)/(98.0 g/mol)
    = 0.015 mol/mL = 15 mmol/mL
    15 M x mL = 0.018 M x 200 mL
    mL conc. H_3PO_4 = 0.24 mL

62. From the Appendix, $K_{a1} = 1.3 \times 10^{-3}$, $K_{a2} = 1.23 \times 10^{-7}$.
    $H_2SO_3 = H^+ + HSO_3^-$ (1)
    $HSO_3^- = H^+ + SO_3^{2-}$ (2)

    The total concentration of sulfurous acid, C_{H2SO3}, is given by
    $C_{H2SO3} = [SO_3^{2-}] + [HSO_3^-] + [H_2SO_3]$ (3)

Using the equilibrium constant expressions to solve for the various sulfurous acid species in terms of $[H_2SO_3]$ and substituting in (3), we reach

$C_{H2SO3} = (K_{a1}K_{a2} [H_2SO_3]/([H^+]^2) + (K_{a1}[H_2SO_3]/([H^+]) + [H_2SO_3]$ from which
$1/\alpha_0 = [C_{H2SO3}]/[H_2SO_3] = K_{a1}K_{a2}/[H^+]^2 + K_{a1}/[H^+] + 1 = 131$
(at pH 4.00)

Similarly,
$1/\alpha_1 = [C_{H2SO3}]/[HSO_3^-] = K_{a2}/[H^+] + 1 + [H^+]/K_{a1} = 1.01$
$1/\alpha_2 = [C_{H2SO3}]/[SO_3^{2-}] = 1 + [H^+]/K_{a2} + [H^+]^2/(K_{a1}K_{a2}) = 820$
$[H_2SO_3] = C_{H2SO3}/(1/\alpha_0) = 0.0100/138 = 7.63 \times 10^{-5} M$
$[HSO_3^-] = C_{H2SO3}/(1/\alpha_1) = 0.0100/1.06 = 9.90 \times 10^{-3} M$
$[SO_3^{2-}] = C_{H2SO3}/(1/\alpha_2) = 0.0100/21 = 1.22 \times 10^{-4} M$

63. $C_{H3PO4} = [H_3PO_3] + [H_2PO_4^-] + [HPO_4^{2-}] + [PO_4^{3-}]$ (1)
Solving first for α_3, substitute in (1) for the various concentrations in terms of $[PO_4^{3-}]$. From the equilibrium constant expressions,

$[HPO_4^{2-}] = ([H^+][PO_4^{3-}])/K_{a3}$ (2)
$[H_2PO_4^-] = [H_3PO_3] = ([H^+]^2[PO_4^{3-}])/(K_{a2}K_{a3})$ (3)
$[H_3PO_3] = ([H^+][H_2PO_4^-])/K_{a1} = ([H^+]^3[PO_4^{3-}])/(K_{a1}K_{a2}K_{a3})$ (4)

Substituting in (1):
$C_{[H3PO3]} = ([H^+]^3[PO_4^{3-}])/(K_{a1}K_{a2}K_{a3}) + ([H^+]^2[PO_4^{3-}])/(K_{a2}K_{a3})$
$\qquad + ([H^+][PO_4^{3-}])/K_{a3} + [PO_4^{3-}]$
which, multiplying by $(1/K_{a1}K_{a2}K_{a3})$ is
$C_{[H3PO3]} = ([H^+]^3[PO_4^{3-}]) + K_{a1} ([H^+]^2[PO_4^{3-}]) + K_{a1}K_{a2} [H^+][PO_4^{3-}]$

$\qquad + K_{a1}K_{a2}K_{a3}[PO_4^{3-}])/(K_{a1}K_{a2}K_{a3})$ (6)
Substituting this in the denominators of the α_3 expression:
$\alpha_3 = [PO_4^{3-}]/C_{[H3PO3]} = (K_{a1}K_{a2}K_{a3})/[H^+]^3 + K_{a1}[H^+]^2 + K_{a1}K_{a2}K_{a3}/[H^+]$
$\qquad + K_{a1}K_{a2}K_{a3})$ (7)

We can take a similar approach for α_1 and α_2, or else we can use (6) as the denominator for α_1 and α_2 and substitute (3) or (2), respectively, for the numerator.

The result gives Equations 7.73 and 7.74.

64. $\mu = 0.100$; $f_{H+} = 0.83$; $f_{CN-} = 0.76$; $K_a^0 = 7.2 \times 10^{-10}$

HCN	=	H$^+$	+	CN$^-$

$\quad 0.0200 - x \qquad x \qquad x$
$\quad \approx 0.0200$
$K_a^0 = ([H^+][CN^-] f_{H+} f_{CN-})/[HCN]$
$7.2 \times 10^{-10} = [(x)(x)(0.83)(0.76)]/0.0200$
$x = [H^+] = 4.8 \times 10^{-6} M$

65. $B + H_2O = BH^+ + OH^-$

$K_b^0 = a_{BH^+} \cdot a_{OH^-}/a_B \approx a_{BH^+} \cdot a_{OH^-}/[B]$

$K_b = [BH^+]f_{BH^+} \cdot [OH^-]f_{OH^-}/[B] = K_b f_{BH^+}f_{OH^-}$

$K_b = K_b/f_{BH^+}f_{OH^-}$

66. The system point is pH = 4.76 at 10^{-3} M HOAc.

This is used as the reference point for the slope = -1 or +1 plots for log [HOAc] or log [OAc⁻]. In strongly acid solution, [HOAc] ≈ 10^{-3} M and in strongly alkaline solution, [OAc⁻] ≈ 10^{-3} M. At pH = pK_a, [HOAc] = [OAc⁻] = 5 x 10^{-4} M, and log [HOAc] = log [OAc⁻] = -3.30. The log [H⁺] and log [OH⁻] curves are as in Figure 7.16.1 (or 7.16.2) in Addendum to Section 7.16 on the text website. Using the above data to plot the curves gives:

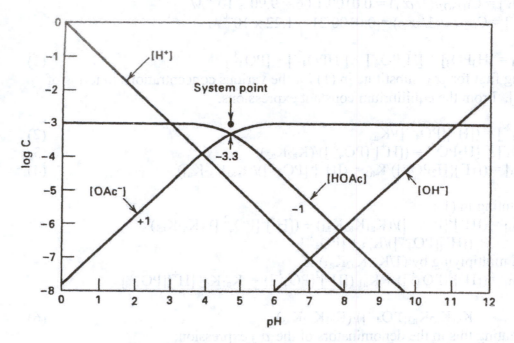

See text website for the spreadsheet setup and chart of the log-log curves.

67. In acid solution, [HOAc] = [OAc⁻]. From the plot where the [H⁺] and [OAc⁻] curves cross, log [H⁺] = -3.88, pH = 3.88. Likewise, [OAc⁻] = $10^{-3.88}$ M

68. log [OAc⁻] = log C$_{HOAc}$K$_a$ + pH

log [OAc⁻] = log (10^{-3})(1.75 x 10^{-5}) + pH

log [OAc⁻] = -7.76 + pH

For pH 2.00:

log [OAc⁻] = -7.76 + 2.00 = -5.76

[OAc⁻] = $10^{-5.76}$ = 1.7 x 10^{-6} M

A similar value is estimated from the log-log diagram.

69. See text website, Chapter 7 Problems for spreadsheet construction of malic acid log-log diagram using alpha values.

70. (a) In malic acid solution, $[H^+] \approx [HA^-]$. This occurs in the graph at $pH \approx 3.33$ (or $\log [H^+] =$ -3.33), and $[HA^-] = 10^{-3.33}$ $M = 4.5 \times 10^{-4}$ M. At this pH, $[A^{2-}] = 10^{-5.00}$ M. (Since $[A^{2-}]$ is $\ll [HA^-]$, we can more accurately calculate the relative $[H_2A] \approx C_{H2A} - [HA^-] = 1.0 \times 10^{-3} - 4.5 \times 10^{-4} = 5.5 \times 10^{-4}$ M, close to the graphical value.)

(b) In a maltate solution, $[OH^-] \approx [HA^-]$, from the hydrolysis of A^{2-}. This occurs at $pH = 8.05$ (or $\log [OH^-] = -5.95$). At this pH, A^{2-} predominates (the hydrolysis is small), and $[HA^-] = 10^{-5.95}$ $M = 1.1 \times 10^{-6}$ M, $[H_2A] = 10^{-7.7}$ $M = 2.0 \times 10^{-8}$ M.

71. $K_{a1} = [H^+][HA^-]/[H_2A]$ $K_{a2} = [H^+][A^{2-}]/[HA^-]$
In very acid solution, $[H_2A] \approx C_{H2A} = 10^{-3}$ M

From K_{a1},
$[HA^-] = K_{a1} C_{HA}/[H^+]$
$\log [HA^-] = \log K_{a1}C_{H2A} + pH = \log(4.0 \times 10^{-4})(10^{-3}) + pH = -6.40 + pH$,
slope $= +1$ for $pH < pK_{a1}$
At pH 1.4, then, $\log [HA^-] = -5$. Check the figure.
In very alkaline solution, $[A^{2-}] = C_{A2-} = 10^{-3}$ M

From K_{a2},
$[HA^-] = C_{HA2}[H^+]/K_{a2}$
$\log [HA^-] = \log(C_{HA2}/K_{a2}) - pH = \log (10^{-3}/8.9 \times 10^{-6}) - pH = 2.05 - pH$,
slope $= -1$ for $pH > pK_{a2}$
At pH 10.05, then, $\log [HA^-] = -8$. Check the figure.

72. (a) At $pH > pK_{a1}$, assume $[H_2PO_4^-] = C_{H3PO4} = 10^{-3}$ M (strictly speaking, this is true half-way between the two pK_a values, i.e., at the first end point of the H_3PO_4 titration).

From K_{a1}, we derive
$\log [H_3PO_4] = \log [H_2PO_4^-]/ K_{a1} - pH$
$\approx \log (10^{-3})/(10^{-1.96}) - pH$
$\log [H_3PO_4] = -1.04 - pH$
At pH 2.96, $\log C = -4.00$. This agrees with the H_3PO_4 curve.

(b) At pH between pK_{a2} and pK_{a3}, assume $[HPO_4^{2-}] = C_{H3PO4} = 10^{-3}$ M (true only at second end point of titration of H_3PO_4, midway between pK_{a2} and pK_{a3}).

From K_{a2}, we derive
$\log [H_2PO_4^-] = \log [HPO_4^{2-}]/ K_{a2} - pH$
$\approx \log (10^{-3})/(10^{-7.12}) - pH$
$\log [H_2PO_4^-] = 4.12 - pH$
At pH 8.12, $\log C = -4.00$, which agrees with the $H_2PO_4^-$ curve.

(c) At pH between pK_{a1} and pK_{a2}, assume $[H_2PO_4^-] = C_{H3PO4} = 10^{-3}$ M.

From K_{a2}, we derive
$\log [HPO_4^{2-}] = \log K_{a2}[H_2PO_4^-] + pH$
$\approx \log (10^{-7.12})(10^{-3}) + pH$
$\log [HPO_4^{2-}] = -10.12 + pH$
At pH 4.12, log C = -6.00, which agrees with the figure.

(d) At pH between pK_{a2} and pK_{a3}, assume $[HPO_4^{2-}] = C_{H3PO4} = 10^{-3}$, as in (b).

(e) At pH between pK_{a2} and pK_{a3}, assume $[HPO_4^{2-}] = C_{H3PO4} = 10^{-3}$, as in (b).

From K_{a3}, we derive
$\log [PO_4^{3-}] = \log K_{a3}[HPO_4^{2-}] + pH = \log (10^{-12.32})(10^{-3}) + pH$
$\log [PO_4^{3-}] = -15.32 + pH$
At pH 10.32, log C = -5.00, which agrees with the curve. Note that these expressions are approximations that hold closest to the midpoints between the pK_a values. The plots begin to curve at the extreme pH values.

73. See the text website for the spreadsheet setup and chart of the H_3PO_4 log-log diagrams using alpha values.

74. The d values given in the website are 4.0-4.5 for Na^+ and $H_2PO_4^-$, 4.0 for HPO_4^{2-}, and 3.0 for K^+. The NIST pH value is 6.833. Input the values for weak acid H_3A. Using d = 4 for Na+ and $H_2PO_4^-$ gives a calculated paH of 6.8246, 0.008 less than the NIST value. Using d = 4.5 for $H_2PO_4^-$ gives paH = 6.8204, 0.013 less than the NIST value. Probably a d of 4.0 for $H_2PO_4^-$ is reasonable, given it is similar to HPO_4^{2-}. Changing the d value of Na^+ to 4.5 does not change the calculation. In any event, the calculated values are within the experimental error of pH measurements (see Section 13.13 of Chapter 13). See the text website Spreadsheet Problem Solutions for a screenshot of the calculation.

75. $K_{a1} = 4.3 \times 10^{-7}$, $pK_1 = 6.37$
$K_{a2} = 4.8 \times 10{-11}$, $pK_2 = 10.32$
$[HCO_3^-] = 0.0250$ M

Inputting for weak acid H_2A gives pH = 8.345. This is the same as calculated in Problem 39. See the text website Spreadsheet Problem Solutions for a screenshot of the calculation.

76. $K_{a1} = 1.1 \times 10^{-2}$; $pK_1 = 1.96$
$K_{a2} = 2.2 \times 10^{-3}$; $pK_2 = 2.66$
$K_{a3} = 6.9 \times 10^{-7}$; $pK_3 = 6.16$
$K_{a4} = 5.5 \times 10^{-11}$; $pK_4 = 10.26$

The concentration of H_3A^- is 0.050 M. Inputting for weak acid H_4A gives pH = 8.209 vs. 8.21 in Problem 41. See the text website Spreadsheet Problem Solutions for a screenshot of the calculation.

77. PFP

The key to this problem is the Henry's law equilibrium equation that is in the problem: If a system is in equilibrium with gaseous CO_2 of any given concentration, regardless of pH or the presence or absence of any other species in the system, has an H_2CO_3 content that is directly governed by the pCO_2 in the gas phase.

The H_2CO_3 content of the water effluent from the soil layer is thus:
$$[H_2CO_3] = K_H p_{CO2} = 4.6 \times 10^{-2} \text{ M/atm} * 3.2 \times 10^{-2} \text{ atm} = 1.4_7 \times 10^{-3} \text{ M}$$

If the total carbonate carbon concentration in the soil effluent (se) water is $C_{T,se}$,
$$[H_2CO_3] = 1.4_7 \times 10^{-3} \text{ M} = C_{T,se}\,\alpha_0$$
$$C_{T,se} = 1.47 \times 10^{-3}/\alpha_0$$
The charge balance equation for a pure H_2CO_3 solution is:
$$[H^+] - K_w/[H^+] - C_{T,se}(\alpha_1 + 2\alpha_2) = [H^+] - K_w/[H^+] - 1.47 \times 10^{-3} \times (\alpha_1 + 2\alpha_2)/\alpha_0 = 0$$
$$[H^+] - K_w/[H^+] - 1.47 \times 10^{-3} \times (K_{a1}[H+] + 2K_{a1}K_{a2})/[H^+]^2 = 0$$
$[H^+]$ is readily computed thence by Goal Seek (See Ch 7 77.xlsx sheet **soil water** in web supplement) to be $8.2_4 \times 10^{-4}$ M thence one can readily calculate that α_0 is essentially 1 and $C_{T,se}$ therefore is 1.47×10^{-3} M, it consists virtually entirely of H_2CO_3.

As this solution percolates through $CaCO_3$, it dissolves $CaCO_3$ until the solubility product equilibrium is reached. No carbonate carbon is lost from the influent water, instead, if s molar $CaCO_3$ dissolves, the the new C_T concentration, C_T', is $C_{T,se} + s$.

The relevant charge balance equation is:
$$2[Ca^{2+}] + [H^+] - K_w/[H^+] - [HCO_3^-] - 2[CO_3^{2-}] = 0$$

We also recognize that $[Ca^{2+}] = s = K_{sp,CaCO3}/[CO_3^{2-}]$; thence:
$$2 \times 4.7 \times 10^{-9}/(\alpha_2 C_T') + [H^+] - K_w/[H^+] - C_T'(\alpha_1 + 2\alpha_2) = 0 \qquad \text{...(7.P77.1)}$$

where
$$C_T' = C_{T,se} + 4.7 \times 10^{-9}/(\alpha_2 C_T') = 1.4_7 \times 10^{-3} + 4.7 \times 10^{-9}/(\alpha_2 C_T')$$
$$C_T'^2 - 1.4_7 \times 10^{-3} C_T' - 4.7 \times 10^{-9}/\alpha_2 = 0$$
or
$$C_T' = 0.5*(1.4_7 \times 10^{-3} + (2.1_6 \times 10^{-6} + 1.8_8 \times 10^{-8}/\alpha_2)^{0.5}) \qquad \text{...(7.P77.2)}$$

We can put equation 7.P77.2 in in 7.P77.1 and the solution is readily obtained by Goal Seek (See Ch 7 77.xlsx sheet **se + CaCO3**). We find that that pH rises significantly upon saturation with $CaCO_3$ (pH 6.87, $[H^+] = 1.35 \times 10^{-7}$) and $4.1_4 \times 10^{-3}$ M $CaCO_3$ dissolves, making $[Ca^{2+}] = 4.1_4 \times 10^{-3}$ M and raising C_T by the same amount from 1.47×10^{-3} M to $5.6_1 \times 10^{-3}$ M.

Imagine now that the solution re-equilibrates with a different gas phase pCO_2, 3.9×10^{-4} atm. This means the new $[H_2CO_3]''$ must be:
$$[H_2CO_3]'' = 4.6 \times 10^{-2} \text{ M/atm} \times 3.9 \times 10^{-4} \text{ atm} = 1.7_9 \times 10^{-5} \text{ M}$$

This will correspond to a new C_T value, C_T'', to $1.7_9 \times 10^{-5}/\alpha_0''$ where the "symbol pertains to this new equilibrium.

We can again write the charge balance equation

$$2[Ca^{2+}] + [H^+] - K_w/[H^+] - [HCO_3^-] - 2[CO_3^{2-}] = 0$$

or

$$2\,K_{sp}/(\alpha_2''C_T'') + [H^+] - K_w/[H^+] - C_T''(\alpha_1'' + 2\alpha_2'') = 0$$

or

$$9.4 \times 10^{-9}/(1.7_9 \times 10^{-5}\,K_{a1}K_{a2}/[H^+]^2) + [H^+] - K_w/[H^+] - 1.7_9 \times 10^{-5}(K_{a1}[H^+] + 2K_{a1}K_{a2})/[H^+]^2 = 0$$

$$5.2_5 \times 10^{-4}\,[H^+]^2/K_{a1}K_{a2} + [H^+] - K_w/[H^+] - 1.7_9 \times 10^{-5}(K_{a1}[H^+] + 2K_{a1}K_{a2})/[H^+]^2 = 0 \quad \ldots(7.P77.3)$$

This equation is also readily solved by Goal Seek. Loss of CO_2 from solution raises pH even further to 8.25 ($[H^+] = 5.69 \times 10^{-9}$ M) and much less calcium can be maintained in solution, the new calcium concentration in solution is $6.0_5 \times 10^{-4}$ M. $4.14 \times 10^{-3} - 6.0_5 \times 10^{-4} = 3.5_4$ mmol $CaCO_3$ per L of the dripping water will end up being deposited as stalactite and stalagmite.

78. PFC
   (i) E; (ii) C; (iii) A; (iv) B; (v) D; (vi) C; (vii) E

Equation A is useful to estimate $[H^+]$ in a weak acid solution.
Equation B is useful to estimate $[H^+]$ in a weak base solution.
Equation C is useful to estimate $[H^+]$ of a salt whose cation has acidic properties (K_a) and anion has basic properties (K_b).
Equation D is useful to estimate the $[H^+]$ of an amphoteric species $H_{n-1}A^-$
Equation E is useful to estimate the $[H_3O^+]$ of an amphoteric species $H_{n-2}A^{2-}$

79. PFP
There are four species, n is 3. The alpha plot shows four curves corresponding to the pH dependent fraction of each of the four species derived from H_3X. The pK's for each of the acid species is given by the pH when $\alpha_n = \alpha_{n+1}$, thus, $pK_1 = 5$, $pK_2 = 7$ and $pK_3 = 12$. This acid will only show one zone of abrupt change because 1) the first two protons are not distinguishable ($pK_2 - pK_1 < 3$), and 2) the HX^{2-} is too weak to be detected with NaOH 0.10 as titrant ($pK + pC_{titrant} > 11$). Since both, titrant and titrated acid, have the same concentration, the only equivalent point occurs at V=30.0 mL (instead of 15.0 mL) because two protons are being titrated.

CHAPTER 8   ACID−BASE TITRATIONS

1. pH = 2. This is the pH range required to change the ratio of [HIN]/[IN⁻] from 1/10 to 10/1, the ratios at which the eye perceives the color of only one form of the indicator.

2. Its transition range must fall within the sharp equivalence point break of the titration curve, i.e., pK_a (In) ≈ pH equivalence point.

3. At pH = pK_a.

4. Alkaline. Because at the end point, we have a solution of a salt of a weak acid, which is a BrØnsted base.

5. For NH_3 vs. HCl, methyl red. For HOAc vs. NaOH, phenolphthalein.

6. CO_2 is boiled out, removing the HCO_3^-/CO_2 buffer system and allowing pH to increase to that of a HCO_3^- solution ([H+] = $\sqrt{K_1 K_2}$). The pH then drops sharply at the end point.

7. pK = 7-8

8. The difference in their K_a's must be > 10^4.

9. A primary standard is one whose purity is known. A secondary standard is one whose purity is unknown and is standardized against a primary standard.

10. A zwitterion forms in solution when an amphoteric substance, such as an amino acid, undergoes proton transfer from the acid group to the basic group.

11. Proteins typically contain 16% nitrogen.

12. Hydrochloric acid is the preferred tritrant because most chlorides are soluble, and it has few side reactions.

13. M_{HCl} = [454.1 mg THAM/121.14 (mg/mmol)]/(35.37 mL) = 0.1060 mmol/mL

14. Reaction: CO_3^{2-} + 2H⁺ → H_2CO_3 ∴ mmol HCl = 2 x mmol Na_2CO_3
    M_{HCl} = [(232.9 mg Na_2CO_3/105.99 mg/mmol x 2 mmol HCl/mmol Na_2CO_3]/42.87mL
    = 0.1025 mmol/mL

15. M_{NaOH} = [859.2 mg KHP/204.23 mg/mmol]/32.67 mL = 0.1288 mmol/mL

16. 1 mmol HCl gives 1 mmol AgCl
    M_{HCl} = = [168.2 mg/143.32 mg/mmol]/143.32 mg/mmol]/10.00 mL
    = 0.1174 mmol/mL

17. $B + H_2O = BH^+ + OH^-$
    $pOH = pK_b + \log [BH^+]/[B]$

    To see acid form:
    $pH = (14 - pK_b) + \log (10/1) = 15 - pK_b$

    To see base form:
    $pH = (14 - pK_b) + \log (1/10) = 13 - pK_b$
    $\Delta pH = (15 - pK_b) - (13 - pK_b) = 2$
    The transition is around $pH = 14 - pK_b$.

18. 0 mL: $pOH = -\log 0.100 = 1.00$; $pH = 13.00$

    10.0 mL: mmol OH^- left $= 0.100\ M \times 50.0$ mL $- 0.200\ M \times 10.0$ mL $= 3.00$ mmol
       $[OH^-] = 3.00$ mmol$/60.0$ mL $= 0.0500\ M$
       $pOH = -\log 0.0500 = 1.30$; $pH = 12.70$

    25.0 mL: All the OH^- has been converted to H_2O. $pH = 7.00$

    30.0 mL: mmol excess $H^+ = 5.0$ mL $\times 0.200\ M = 1.0$ mmol$/80$ mL $= 0.012_5\ M$
       $pH = -\log 0.012_5 = 1.90$

19. 0 mL: $[H^+] = \sqrt{K_a[HA]} = \sqrt{2.0 x 10^{-5} x 0.200} = 2.0 \times 10^{-3}\ M$
       $pH = 2.70$

    10.0 mL: mmol A^- produced $= 0.100$ M $\times 10.0$ mL $= 1.00$ mmol
       mmol HA left $= 0.200\ M \times 25.0$ mL $- 1.00$ mmol $= 4.00$ mmol
       $pH = -\log 2.5 \times 10^{-5} + \log (1.00/4.00) = 4.10$

    25.0 mL: mmol A^- formed $= 0.100\ M \times 25.0$ mL $= 2.50$ mmol
       mmol HA left $= 5.00\ M \times 25.0$ mL $= 2.50$ mmol
       $pH = 4.70 + \log (2.50/2.50) = 4.70$

    50.0 mL: mmol A^- formed $= 0.100\ M \times 50.0$ mL $= 5.00$ mmol in 75.0 mL
       All the HA has been converted to A^-. $[A^-] = 0.0667\ M$
       $[OH^-] = \sqrt{[(1.0 x 10^{-14})/(2.0 x 10^{-5})] x 0.0667} = 5.77 \times 10^{-6}\ M$
       $pOH = 5.24$; $pH = 8.76$

    60.0 mL: mmol excess $OH^- = 0.100\ M \times 10.0$ mL $= 1.00$ mmol in 85.0 mL
       $[OH^-] = 0.0118\ M$
       $pOH = 1.93$; $pH = 12.07$

20. 0 mL: $[OH^-] = \sqrt{K_b[NH_3]} = \sqrt{1.75x10^{-5}x0.100} = 1.32 \times 10^{-3}\ M$
    pOH = 2.88; pH = 11.12

   10.0 mL: mmol NH_4^+ formed = 0.100 M x 10.0 mL = 1.00 mmol
   mmol NH_3 left = 0.100 M x 50.0 mL – 1.00 mmol = 4.00 mmol
   pH = 14.00 – [-log 1.75 x 10^{-5} + log (1.00/(4.00))] = 9.84

   25.0 mL: mmol NH_4^+ formed = 0.100 M x 25.0 mL = 2.50 mmol
   mmol NH_3 left = 5.00 – 2.50 = 2.50 mmol
   pOH = 4.76 + log (2.50/2.50) = 4.76; pH = 9.24

   50.0 mL: mmol NH_4^+ formed = 0.100 M x 50.0 mL = 5.00 mmol in 100 mL. All the
   NH_3 has been converted to NH_4^+. $[NH_4^+]$ = 0.0500 M.
   $[H^+] = \sqrt{[(1.0x10^{-14})/(1.75x10^{-5})]x0.0500} = 5.35 \times 10^{-6}\ M$
   pH = 5.47

   60.0 mL: mmol excess H^+ = 0.100 M x 10.0 mL = 1.00 mmol in 110 mL
   $[H^+]$ = 0.0909 M; pH = 2.04

21. 0%: Calculate pH from ionization of first proton of H_2A. Since [HA] is just equal to 100 x
   K_{a1}, the quadratic equation need not be used.
   $[H^+] = \sqrt{K_{a1}x[H_2A]} = \sqrt{1.0x10^{-3}x0.100} = 1.0 \times 10^{-2}\ M$
   pH = 2.00

   25.0% (50 mL): This is halfway to the first equivalence point and $[H_2A] = [HA^-]$
   (0.100 M x 50.0 mL = 5.00 mmol each)
   pH = pK_{a1} + log $[HA^-]/[H_2A]$ = 3.00 + log (5.00/5.00) = 3.00

   50.0% (100 mL): This is at the 1st equivalence point. All H_2A has been converted to HA$^-$
   (= 10.0 mmol in 200 mL).
   $[H^+] = \sqrt{K_{a1}K_{a2}} = \sqrt{1.0x10^{-3}x1.0x10^{-7}} = 1.0 \times 10^{-5}\ M$
   pH = 5.00

   75.0% (150 mL): This is halfway between the two equivalence points, and $[HA^-] = [A^{2-}]$
   (5.00 mmol each)
   pH = pK_{a2} + log $([A^{2-}]/[HA^-])$ = 7.00 + log (5.00/5.00) = 7.00

   100.0% (200 mL): All the HA$^-$ has been converted to A^{2-} (= 10.0 mmol in 300 mL)
   $[A^{2-}]$ = 0.0333 M
   $A^{2-} + H_2O = HA^- + OH^-$
   $[HA^-][OH^-]/[A^{2-}] = (K_w/K_{a2}) = (1.0 \times 10^{-14}/1.0 \times 10^{-7}) = 1.0 \times 10^{-7}$
   $[OH^-] = \sqrt{1.0x10^{-7}x[A^{2-}]} = \sqrt{1.0x10^{-7}x0.0333} = 5.8 \times 10^{-5}\ M$
   pOH = 4.24; pH = 9.76

125%: mmol excess OH^- = 0.100 M x 50 mL = 5.0 mmol in 350 mL
   (250 mL)
   (The hydrolysis of A^{2-} is negligible in the presence of added OH^-.)
   $[OH^-]$ = 0.0143 M
   pOH = 1.84; pH = 12.16

22. 0%: $[H^+] = \sqrt{K_{a2} x K_{a3}} = \sqrt{(7.5x10^{-8})(4.8x10^{-13})} = 1.9_0 \times 10^{-10}$ M

   25.0%: mmol $H_2PO_4^-$ formed = 0.100 M x 25.0 mL = 2.50 mmol
      mmol HPO_4^{2-} left = 0.100 M x 100 mL - 2.50 mmol = 7.50 mmol
      pH = pK_{a2} + log $[HPO_4^{2-}]$/$[$ $H_2PO_4^-]$ = 7.12 + log (7.50/2.50) = 7.60

   50.0%: mmol $H_2PO_4^-$ formed = 0.100 M x 50.0 mL = 5.00 mmol
      mmol HPO_4^{2-} left = 10.0 mmol - 5.00 mmol = 5.0 mmol
      pH = pK_{a2} = 7.12

   75.0%: mmol $H_2PO_4^-$ formed = 0.100 M x 75.0 mL = 7.50 mmol
      mmol HPO_4^{2-} left = 10.0 – 7.50 = 2.5 mmol
      pH = 7.12 + log (2.50/7.50) = 6.64

   100.0%: mmol $H_2PO_4^-$ formed = 0.100 M x 100 mL = 10.0 mmol in 200 mL
      All the HPO_4^{2-} has been converted to $H_2PO_4^-$. $[$ $H_2PO_4^-]$ = 0.0500 M
      $[H^+] = \sqrt{K_{a1} x K_{a2}} = \sqrt{(1.1x10^{-2})(7.5x10^{-8})}$ = 2.8 x 10^{-5} M
      pH = 4.54

   150%: mmol H_3PO_4 formed = 50.0 mL x 0.100 M = 5.00 mmol
      mmol $H_2PO_4^-$ left = 10.0 – 5.00 = 5.0 mmol
      pH = pK_{a1} + log $[H_2PO_4^-]$/$[$ $H_3PO_4]$ = 1.96 + log (5.0/5.0) = 1.96

23. % KH_2PO_4 = {[25.6 mL x 0.112 (mmol/mL) x 136 (mg/mmol)]/(492 mg)} x 100%
   = 79.2%

24. 90.0% = {[0.155 (mmol/mL) x x mL x 2 (mmol LiOH/mmol H_2SO_4)
   x 23.95 (mg/mmol)]/(293 mg)} x 100%
   x = 35.5 mL

25. [0.250 (mmol/mL) x 25.0 mL – 0.250 (mmol/mL) x 9.26 mL] x 56.1 mg/mmol
   = 221 mg KOH reacted. Saponification no. = 221 mg KOH/1.10 g fat = 201 mg/g
   {[3.94 mmol KOH reacted x 1/3 (mmol fat/mmol KOH) x f.w._fat]/(1100 mg)}
   x 100% = 100% fat
   f.w._fat = 838 mg/mmol

26. mmol alanine = mmol N = 2 mmol H_2SO_4 reacted

∴ % alanine = {[mmol H_2SO_4 – mmol NaOH x ½) x 2 x $CH_3CH(NH_2)COOH$]/(mg sample)}
x 100%

= {[(0.150 x 50.0 – 0.100 x 9.0 x ½) x 2 x 89.10]/(2000)} x 100% = 62.8%

27. Standardization: net reaction is

$(NH_4)_2SO_4$ + 2 HCl = 2 NH_4Cl + H_2SO_4

∴ mmol HCl = 2 x mmol $(NH_4)_2SO_4$

M_{HCl} x 33.3 mL = 2 x [(330 mg)/(132.1 mg/mmol)]

M_{HCl} = 0.150 M

g protein/100 mL =

[(0.150 M x 15.0 mL x 0.01401 g N/mmol x 6.25 g protein/g N)/(2.00 mL)] x 100 mL

= 9.85 g %

28. mmol H_3PO_4 = 0.200 M x 10.0 mL = 2.00 mmol

mmol HCl = 0.200 M (25.0 – 10.0) mL = 3.00 mmol

∴ H_3PO_4 = 2.00 mmol/100 mL = 0.0200 M

HCl = 3.00 mmol/100 mL = 0.0300 M

29. The first end point corresponds to adding 1 H^+ to CO_3^{2-} and the second end point (28.1 mL later) corresponds to adding 1 more 1 H^+ to CO_3^{2-} plus titration of the original HCO_3^-.

mmol CO_3^{2-} = 0.109 M x 15.7 mL = 1.71 mmol

mmol HCO_3^- = 0.109 M (28.1 – 15.7) mL = 1.35 mmol

% Na_2CO_3 = [(1.71 mmol x 106.0 mg/mmol)/(527 mg)] x 100% = 34.4%

% $NaHCO_3$ = [(1.35 mmol x 84.0 mg/mmol)/(527 mg)] x 100% = 21.5%

30. mg Na_2CO_3 = 0.250 M x 15.2 mL x 106 mg/mmol = 403 mg

mg NaOH = 0.250 M x (26.2 – 15.2) mL x 40.0 mg/mmol = 110 mg

31. M_{HCl} = [(477 mg Na_2CO_3)/106.0 mg/mmol) x 2 (mmol HCl/mmol Na_2CO_3)]
/(30.0 mL) = 0.300 mmol/mL

The first end point is shorter than the second and corresponds to titration of CO_3^{2-} to only HCO_3^-. The second end point corresponds to titration of CO_3^{2-} (in the form of HCO_3^-) plus HCO_3^- originally present to H_2CO_3.

mmol Na_2CO_3 = 15.0 mL x 0.300 M x 1 (mmol Na_2CO_3/mmol HCl) = 4.50 mmol

The same volume is used to titrate the CO_3^{2-} to the second end point:

mmol $NaHCO_3$ = (35.0 – 15.0) mL x 0.300 M x 1 (mmol $NaHCO_3$/mmol HCl)
= 6.00 mmol

32. In this case, the first end point is longer than the second and corresponds to titration of CO_3^{2-} to HCO_3^- plus OH^- to H_2O. The second end point corresponds to titration of only CO_3^{2-} (in the form of HCO_3^-) to H_2CO_3.

mmol Na_2CO_3 = 10.0 mL x 0.300 M x 1 (mmol Na_2CO_3/mmol HCl) = 3.00 mmol

mmol NaOH = (15.0 – 10.0) mL x 0.300 M x 1 (mmol NaOH/mmol HCl)

= 1.5_0 mmol

33. Let x = mg $BaCO_3$. Then mg Li_2CO_3 = 150 – x

$mmol_{HCl}$ = $mmol_{BaCO3}$ x 2 + $mmol_{Li2CO3}$ x 2

= $[mg_{BaCO3}/f.w_{BaCO3}$ (mg/mmol)$]$ x 2 (mmol HCl/mmol $BaCO_3$)

+ $[mg_{Li2CO3}/$ f.w.$_{Li2CO3}$ (mg/mmol)$]$ x 2 mmol HCl/mmole Li_2CO_3

0.120 M (mmol/mL) x 25.0 mL = (x mg_{BaCO3}/197.35 mg/mmol) x2

+ $[(150 - x)$ mg_{Li2CO3}/73.89 mg/mmol$]$ x 2

x = 62.6 mg $BaCO_3$

% $BaCO_3$ = (62.6 mg)/(150 mg) x 100% = 41.7%

34. Let x = mg H_3PO_4. Then mg P_2O_5 = 405 – x

$mmol_{NaOH}$ = $mmol_{H3PO4}$ x 2 (mmol NaOH/mmol H_3PO_4) + mmol P_2O_5 x 4 (mmol NaOH/mmol P_2O_5)

= $[mg_{H3PO4}/f.w._{H3PO4}$ (mg/mmol)$]$ x 2 + $[mg_{P2O5}/f.w._{P2O5}$ (mg/mmol)$]$ x 4

0.250 M (mmol/mL) x 42.4 mL = $[x$ mg_{H3PO4}/98.0 mg/mmol)$]$ x2 + $[(405 - x)$ mg_{P2O5}/141.9 (mg/mmol)$]$ x 4

x = 10_2 mg H_3PO_4

% H_3PO_4 = (10_2 mg/405 mg) x 100% = $25._2$

35. Since NaOH is a strong base, pH 12 means 0.01 M NaOH. 100 mL amounts to 0.01 mole/L x 0.1 L = 1 mmol.

(a) drop in pH after one hour:

    Total volume of air filtered in one hour = 5x60 = 300 mL.

    Total volume of CO_2 passing in one hour = 0.5*300/100 = 1.5 mL.

    From: PV = Nrt

    Total amount of CO_2 sequestered in one hour ≡ n = PV/RT = 1x($1.5x10^{-3}$)/(0.082x298) = $6.14x10^{-5}$ moles.

    From the stoichiometric equation: $2OH^- + CO_2(g) \rightarrow CO_3^{2-} + H_2O$

    Total amount of OH^- consumed in one hour = $6.14x10^{-5}$x2/1 = $1.28x10^{-4}$ moles.

    Amount of OH^- remaining after one hour = 1-0.128 = 0.872 mmol

    Concentration of OH^- = 0.872/100 mmol/mL = 8.72 x 10^{-3} M

    pOH = -log($8.72x10^{-3}$) = 2.06, pH after one hour = 14-pOH = 11.94

(b) drop in pH after 5 hours, Repeat the same set of calculations where, $[OH^-]$ = $3.86x10^{-3}$ M, pOH = 2.41, pH after 5 hours = 11.59

36. It's a monoprotic acid and the pH will increase with the titration, being about 1.9 (pK_a) at the midpoint, so (C).

37. (b) The first equivalent point corresponds to the titration of H_2A first proton; the second equivalent point corresponds to the titration of the original HA- species plus the HA- resulting from the H_2A titration. Thus, the second EP requires twice as much volume as the first EP.

38. (b) The feasibility of a weak acid titration with strong base depends on both the acid strength and the titrant concentration. As a rule of thumb, for the titration of a weak acid with strong base, an EP is observed whenever $C_{base}K_a > 10^{-11}$. Thus, by increasing the C_{base}, it is possible to improve the equivalence point detection of a given weak acid.

Explanation: By increasing C_{base}, the pH upper limit also increases while keeping the pH of the buffer region unaltered.
To observe the plot at right, both acid and base must have equal concentrations (Equivalence point 25 mL as in the plot at left) and higher than 0.01 M.

39. Pre-equivalence region: For the weak acid titration, the pH depends on the acid/conjugate base ratio at the so-called buffer region, while for the strong acid titration, it depends on the non-titrated acid which remains in solution. By diluting the weak acid, only the pH near 0 mL is a little higher, as it varies with the square root of C_a, however, in the buffer region the pH remains essentially the same as for the more concentrated acid because the ratio of conjugate base to acid is also the same. In the case of the strong acid, since the pH depends on the concentration of the remaining acid, by diluting it, the pH increases.

NOTE: If the weak acid is extensively diluted, it eventually starts behaving as a strong acid!!

Post-equivalence region: In both cases the pH depends on the water autoprotolysis, influenced by the excess NaOH.

40. The key to solving the problem is to compare volumes at the equivalence points within each plot. In plot A, $V_1 = 20$ mL and $V_2 = 50$ mL. The second equivalence point required more titrant (30 mL) than the first equivalence point (20 mL). This is compatible with having a mixture of CO_3^{2-} / HCO_3^{-}. For the first equivalence point you are titrating CO_3^{2-} to HCO_3^{-}. For the second equivalence point, you are titrating both, the original HCO_3^{-} and the one coming from CO_3^{2-}. To calculate the composition of the mixture, then
mmole $CO_3^{2-} = M_{HCl}V_1$
mmole $HCO_3^{-} = M_{HCl}V_2 - 2 M_{HCl}V_1$

$$M_{CO_3^{2-}} = \frac{0.010(20)}{10} = 0.020M$$

$$M_{HCO_3^{-}} = \frac{0.010(50) - 2(0.010(20))}{10} = 0.010M$$

In plot B, $V_1 = 25$ mL; $V_2 = 40$ mL. It is not possible for this mixture to be CO_3^{2-} and HCO_3^{-} because the volume needed to reach the second equivalence point from the first equivalence point (15 mL) is less than V_1. In this case the first 25 mL are used to titrate both OH^- to H_2O and CO_3^{2-} to HCO_3^{-}. The next 15 mL (40 mL - 25 mL) are used to titrate HCO_3^{-} to H_2CO_3. To calculate the composition of the mixture, then
mmole $CO_3^{2-} = M_{HCl}(V_2-V_1)$
mmole $OH^- = M_{HCl}(V_1-(V_2-V_1)) = M_{HCl}(2V_1 - V_2)$

$$M_{CO_3^{2-}} = \frac{0.010(40-25)}{10} = 0.015M$$

$$M_{OH^-} = \frac{0.010(2 \times 25 - 40)}{10} = 0.010M$$

41. (b). See the inside back cover of the book for the behavior of phenol red. The phenol red indicator pH range is 6.4 – 8.2. This range corresponds to the second equivalence point:

$H_3A + 2\,OH^- \rightarrow HA^- + 2\,H_2O$

$$molar\ mass = \frac{mg\ of\ H_3A\ in\ 25\ mL\ aliquot}{mmole\ of\ H_3A\ in\ 25\ mL\ aliquot}$$

$$= \frac{\left(\dfrac{25.00\ mL}{100.0\ mL}\right) \times 400.0}{15.90\ mL \times 0.0750\ mmole/mL\left(\dfrac{1\,mmole\ H_3A}{2\ mmole\ OH^-}\right)} = 167.7$$

42. (e). An H_2A titration with 0.100 M NaOH will show just one equivalence point with 2:1 stoichiometry if $pK_2 - pK_1 < 3$ and $pK_2 \leq 10$. Only tartaric acid meets those criteria.

43. (c) An H_3A titration with 0.100 M NaOH will show three equivalence points if $pK_n - pK_{n+1} \geq 3$ and $pK_3 \leq 10$. Only histidine meets those criteria.

44. The formula weight of KHP is 204.221 g/mol
[(0.1037 M × 0.02823 L × 204.221 g/mol)/0.9872 g] × 100 % = 60.56%

45. $CaCO_3 + 2HCl \rightarrow CaCl_2 + CO_2 + H_2O$
$NaOH + HCl \rightarrow NaCl + H_2O$
The formula weight of $CaCO_3$ is 100.086 g/mol
[(0.1035 M × 0.05000 L – 0.1018 M × 0.01662 L)/2] × 100.086 g/mol = 0.1743 g
(0.1743 g/0.2027 g) × 100 = 86.00%

46. Total HCl consumed = NaOH + NH_3
5.00 mL x 0.03000 M = (10.00 mL x 0.010) + NH_3
NH_3 = 0.0500 mmol
0.0500 mmol NH_3, the weight of nitrogen is:
0.0500 mmol ´ 14.01 mg/mmol = 0.7005 mg

The weight percent of nitrogen in the protein = $\dfrac{0.7005\,mg}{0.500\,mL \times 50.0\,mg/mL} \times 100\% = 2.80\%$

47. Phosphoric acid - 97.995 g/mol $K_{a1} = 1.1 \times 10^{-2}$ $K_{a2} = 7.5 \times 10^{-8}$ $K_{a3} = 4.8 \times 10^{-13}$
    Citric acid - 192.124 g/mol $K_{a1} = 7.4 \times 10^{-4}$ $K_{a2} = 1.7 \times 10^{-5}$ $K_{a3} = 4.0 \times 10^{-7}$
    Sodium Hydroxide - 39.997 g/mol

    (a) For 1 L of solution: 12.00 mg/L H_3PO_4 or 0.01200 g/L = (0.01200 g/L)/(97.995 g/mol) = 1.224×10^{-4} mol/L H_3PO_4

Phosphoric acid:
$K_{a1} = 1.1 \times 10^{-2}$
$K_{a2} = 7.5 \times 10^{-8}$
$K_{a3} = 4.8 \times 10^{-13}$

Using a charge balance equation and Goal Seek (see 8.48.xlsx in the book's website) we obtain $[H^+] = 1.21_1 \times 10^{-4}$, pH = 3.917. Inputting the same data into the Stig Johansson program for ConcpH also gives the same pH.

(b) To get $2 \times 1.224 \times 10^{-4}$ mol citric acid, fw 192.124 g/mol, we will need
= $2 \times 1.224 \times 10^{-4} \times 192.124 = 4.704 \times 10^{-2}$ g = 47.04 mg

(c) Using a charge balance equation, solving the same problem with twice the same total concentration of acid but with K_a values for citric acid and with a sodium term incorporated, Goal Seek to find the $[Na^+]$ produces $[Na^+] = 1.201 \times 10^{-4}$ M. Multiplying by fw NaOH (39.997 g/mol), we need 4.80 mg.

This amount is very small and will actually be difficult to weigh accurately especially because of the great affinity of NaOH for moisture and CO_2.

48. The pH of the sulfite-hydrogensulfite buffer is readily computed from the Henderson-Hasselbalch equation: pH = 6.91 + log (0.090/0.030)   pH = 7.39

The total concentration is 0.09+0.03 = 0.12 M.
The only buffering system that has a pK_a with one unit of this pH is the pK_a of $H_2PO_4^-$, at 7.2.
Let us consider making 1 L buffer. If we take 120 mmol of Na_2HPO_4 (141.949 g/mol), we need
$0.12 \times 141.949 = 17.034$ g. We will need to generate x amounts of PO_4^{3-} out of this, such that
$7.39 = 7.20 + \log (x/(0.12-x))x/(0.12-x) = 10^{0.19} = 1.55$
x = 0.186 – 1.55x
1.65 x = 0.186
x = 0.0729 mole
72.9 mmol NaOH x 39.997 mg/mmol = 2.916 g NaOH

49. pH 2.45 for a strong monoprotic acid like HNO_3 indicates a molar concentration of $10^{-2.45}$ = 3.548×10^{-3} mol/L (M).
25 wt% HNO_3 of density 1.18: 25 g HNO_3/100g x 1.18g/mL → 250 g/1,000g x1,180g/L = 295 g HNO_3/L → 295/63.013 = 4.68 M
Using $M_1V_1 = M_2V_2$
$500 \times 3.548 \times 10^{-3} = 4.68 \times V_2$; $V_2 = 0.379$ mL

50. (a) From Henderson-Hasselbalch Equation, pH=6.50.

(b) You would expect it to be lower. That would be true at any volume, because it will take a smaller amount of acid (or base) to change the pH by a given amount.

(c) Until the concentration is dropped to low (perhaps 10^{-4} M or lower), not much. On the assumption that we are dealing with buffers, it makes no sense to use low concentrations.

(d) Don't ever eat or drink anything found in the laboratory, even if the label says that it should be harmless, sodium chloride, for example.

51. Take the volume of 0.100 M Na_2HPO_4 you want to start with. Also make a solution of 0.100 M NaH_2PO_4. Add the latter to the former in small portions while measuring the pH and adjust to pH 7.0. You can also use 0.100 M H_3PO_4 but it is easier to make NaH_2PO_4 in exactly known concentration as it can be weighed out.

52. (a) with a given $[H^+]$ value we can directly calculate the value of $[OH^-]$ and all the relevant α values (and thence all individual anion concentrations), it is therefore readily possible to calculate the needed $[Na^+]$ concentration by difference. On the other hand, when $[Na^+]$ is given instead, a complex polynomial must be solved for which we use Goal Seek or Solver.

(b) If we have a pH value which is below what would be attained by the pure acid, the calculated added base volume will be negative. This is tantamount to saying that amount of strong acid, the same concentration as the base, is needed.

(c,d) Check yourself!

53. The general approach that Professor Goates has in the KHP titration spreadsheet (8.53.54.xlsx) is very similar to problem 52 in that the amount of base added is what is calculated for a given pH. Of course the initial pH has to be guessed, if you guess it to be too low, you will end up with a negative value for the needed base, as we observed in problem 52.

To begin, to V_A^0 mL C_A^0 M KHP we added V_w mL water such that the initial volume is V_A mL ($= V_A^0 + V_w$) and the initial concentration is C_A ($=V_A^0 C_A^0/(V_A^0 + V_w)$). Charge balance requires that:

$$[Na^+] + [K^+] + [H^+] - [OH^-] - (\alpha_1 + 2\alpha_2)C_A = 0$$

Transposing all but $[Na^+]$

$$[Na^+] = [OH^-] - [H^+] + (\alpha_1 + 2\alpha_2)C_A - [K^+] = 0$$

For KHP, $[K^+] = C_A$ as well so we can write

$$[Na^+] = [OH^-] - [H^+] + (\alpha_1 + 2\alpha_2 - 1)C_A$$

Let us assume that V_B mL of base of molarity C_B is added, the total volume is now $V_A + V_B$, thus:

$$C_B V_B/(V_A+V_B) = [OH^-] - [H^+] + (\alpha_1 + 2\alpha_2 - 1)C_A V_A/(V_A+V_B)$$

Multiplying throughout by (V_A+V_B)

$$C_B V_B = ([OH^-] - [H^+])(V_A+V_B) + (\alpha_1 + 2\alpha_2 - 1)C_A V_A$$

$$C_B V_B + ([H^+] - [OH^-]) V_B = ([OH^-] - [H^+])V_A + (\alpha_1 + 2\alpha_2 - 1)C_A V_A$$

$$V_B (C_B + [H^+] - [OH^-]) = V_A ([OH^-] - [H^+] + (\alpha_1 + 2\alpha_2 - 1)C_A)$$

$$V_B = \{V_A ([OH^-] - [H^+] + (\alpha_1 + 2\alpha_2 - 1)C_A)\}/(C_B + [H^+] - [OH^-])$$

In the spreadsheet 8.53.54.xlsx, on worksheet 53, the denominator and the numerator are separately calculated, respectively, as:

$$\text{Denominator} = C_B + [H^+] - [OH^-]$$

$$\text{Numerator} = V_A ([OH^-] - [H^+] + (\alpha_1 + 2\alpha_2 - 1)C_A)$$

And hence V_B is calculated from the ratio.

In worksheet "Effect of V_w" and the plot "53 Tit'n Plot 2" the calculations are done for $V_w = 6$ mL instead of 5 mL and both data are plotted together.

54. The first portion of the worksheet up to column G is exactly the same as that in 53. In preparation to calculating ionic strength μ we calculate $[Na^+]$ in column H as $V_B C_B/(V_A+V_B)$, $[K^+]$ in column I as $V_A C_A/(V_A+V_B)$, [HP-] in column J as $[K^+]\alpha_1$, and $[P^{2-}]$ in column K as $[K^+]\alpha_2$. The ionic strength μ is then computed as $\frac{1}{2}$ $([H^+] + [OH^-] + [Na^+] + [K^+] + [HP^-] +4*[P^{2-}])$. Then the activity coefficient f_i is computed as $10^{\wedge}(-0.51*\text{SQRT}(\mu)/(1+ 0.33a_i*\text{SQRT}(\mu)))$ where a_i is the ion size parameter in columns M, N, O, P for H^+, HP^-, P^{2-}, and OH^-, respectively. Concentration based K_{a1}, K_{a2} and K_w are then respectively computed in columns Q, R and S as $K_{a1}/(f_{H+}*f_{HP-})$, $K_{a2}*f_{HP-}/(f_{H+}*f_{P2-})$, and $K_w/(f_{H+}*f_{OH-})$. In columns T – W α_1, α_2, V_B and thence μ is recalculated. This entire process up to the calculation of V_B is then repeated in columns X to AG. The newly calculated V_B values vs. pH (C vs AG) is then replotted with the data obtained previously without activity corrections

CHAPTER 9   COMPLEXOMETRIC REACTIONS AND TITRATIONS

1. A chelating agent is a type of complexing agent in which the complexing molecule has two or more complexing groups.

2. Chelation titration indicators are themselves chelating agents that form a weaker chelate with the titrate metal ion than does the titrant. The titrant displaces all the indicator from the metal ion at the equivalence point, causing the color to revert to that of the uncomplexed indicator.

3. The indicator forms too weak a chelate with calcium to give a sharp end point. The added magnesium combines with the indicator, and the end point occurs when the EDTA displaces the last of the indicator from the magnesium. The EDTA is standardized after adding the magnesium in order to correct for the decrease in the effective molarity of EDTA due to reaction with magnesium.

4. Because the ammonia complexes the copper, which results in a small conditional formation constant for the Cu-EDTA complex, and hence a small endpoint break. See Example 9.6.

5. After mixing, we start with 0.0500 M Ca^{2+} and 1.0 M NO_3^-. We neglected the amount of reacted NO_3^-. At equilibrium we have:

$$Ca^{2+} \;+\; NO_3^- \;=\; Ca(NO_3)^+$$
$$x \qquad 1.0 - x \qquad 0.0050 - x$$

Try neglecting x compared with 1.0, but not compared to 0.0050.
$[Ca(NO_3)^+]/[Ca^{2+}][NO_3^-] = K_f = 2.0$
$(0.0050 - x)/[(x)(1.0)] = 2.0$
$x = [Ca^{2+}] = 0.0017\ M$
$[Ca(NO_3)^+] = 0.0050 - 0.0017 = 0.0033\ M$

6. Let en = $NH_2CH_2CH_2NH_2$

$$en \;+\; Ag^+ \;=\; Ag(en)^+$$
$$x \qquad x \qquad 0.10 - x$$

$[Ag(en)^+]/[en][Ag^+] = 5.0 \times 10^4$
Assume x is negligible compared to 0.10. Then,
$(0.10)/(x)(x) = 5.0 \times 10^4$
$\quad x = 1.4 \times 10^{-3}\ M = [Ag^+]$

7. $\qquad en \;+\; Ag^+ \;=\; Ag(en)^+$
$\quad 0.10 + x \qquad x \qquad 0.10 - x$

Again, $x \ll 0.10$, so
$(0.10)/[(0.10)(s)] = 5.0 \times 10^4$
$x = 2.0 \times 10^{-5} M = [Ag^+]$

8. This problem is a simpler version of Example 9.7.
    We are given $K_{f1} = 6.6 \times 10^8$ and $K_{f2} = 4.4 \times 10^4$. Therefore, $\beta_1 = K_{f1} = 6.6 \times 10^8$
    $\beta_2 = K_{f1} \times K_{f2} = 6.6 \times 10^8 \times 4.4 \times 10^4 = 2.9 \times 10^{13}$
    According to 9.19, $\alpha_{Ag} = (1 + \beta_1 L + \beta_2 L^2)^{-1}$

    Thiosulfate, $S_2O_3^{2-}$ is the ligand L; L = 1.00. This is so much larger than $C_{Ag} = 0.0100$,   the
    consumption of the $S_2O_3^{2-}$ due to complexation will result in a negligible change in the
    concentration of free $S_2O_3^{2-}$. Thus,
    $\alpha_{Ag} = (1 + \beta_1 L + \beta_2 L^2)^{-1} = (1 + 6.6 \times 10^8 \times 1.00 + 2.9 \times 10^{13} \times 1.00^2)^{-1} = 3.4 \times 10^{-14}$
    $[Ag^+] = C_{Ag} \times \alpha_{Ag} = 0.0100 \times 3.4 \times 10^{-14} = 3.4 \times 10^{-16}$ M
    $[AgL^-] = \beta_1 \times [Ag^+] \times L = 6.6 \times 10^8 \times 3.4 \times 10^{-16} \times 1.00 = 2.2 \times 10^{-7}$ M
    $[AgL_2^{3-}] = \beta_2 \times [Ag^+] \times L^2 = 2.9 \times 10^{13} \times 3.4 \times 10^{-16} \times 1.00^2 = 9.9 \times 10^{-3}$ M

9. (a) From Equation 9.12,
       $1/\alpha_4 = 1 + (10^{-3})/(5.5 \times 10^{-11}) + [(10^{-3})^2]/[(6.9 \times 10^{-7})(5.5 \times 10^{-11})] + [(10^{-3})^3]/[(2.2 \times 10^{-3})$
       $(6.9 \times 10^{-7})(5.5 \times 10^{-11})] + [(10^{-3})^4]/[(1.0 \times 10^{-2})(2.2 \times 10^{-3})(6.9 \times 10^{-7})(5.5 \times 10^{-11})]$
       $\alpha_4 = 2.5_3 \times 10^{-11}$
       $K_f' = K_f \alpha_4 = 1.10 \times 10^{18} \times 2.5_3 \times 10^{-11} = 2.7_8 \times 10^7$

    (b) $1/\alpha_4 = 1 + (10^{-10})/(5.5 \times 10^{-11}) + [(10^{-10})^2]/[(6.9 \times 10^{-7})(5.5 \times 10^{-11})]$
       $+ [(10^{-10})^3]/[(2.2 \times 10^{-3})(6.9 \times 10^{-7})(5.5 \times 10^{-11})]$
       $+ [(10^{-10})^4]/[(1.0 \times 10^{-2})(2.2 \times 10^{-3})(6.9 \times 10^{-7})(5.5 \times 10^{-11})]$
       $\alpha_4 = 0.35_5$
       $K_f' = K_f \alpha_4 = 1.10 \times 10^{18} \times 0.35_5 = 3.9_0 \times 10^{17}$

10. (a) $K_f' = 2.7_8 \times 10^7$
       (1) $pPb = -\log [Pb^{2+}] = 1.60$

       (2) mmol Pb^{2+} start $= 0.0250 M \times 50.0$ mL $= 1.25$ mmol
           mmol EDTA added $= 0.0100 M \times 50.0$ mL $= 0.50$ mmol
           $=$ mmol PbY^{2-} formed
           mmol Pb^{2+} left $= 0.75$ mmol

       Neglecting dissociation of PbY^{2-}, $pPb = -\log (0.75$ mmol$/100$ mL$) = 2.12$

       (3) Stoichiometric amounts of Pb^{2+} and EDTA $=$
           1.25 mmol $PbY^{2-}/175$ mL $= 7.14 \times 10^{-3} M$
           $[PbY^{2-}]/[Pb^{2+}]C_{H4Y} = 2.7_8 \times 10^7$
           $(7.14 \times 10^{-3} - x)/[(x)(x)] = 2.7_8 \times 10^7 \approx (7.14 \times 10^{-3})/x^2$
           $x = 1.60 \times 10^{-5} M = [Pb^{2+}]$; $pPb = -\log 1.60 \times 10^{-5} = 4.80$

(4) $[PbY^{2-}] = 1.25$ mmol/250 mL $= 5.00 \times 10^{-3}\ M$
mmol excess EDTA $= 0.0100\ M \times 200$ mL $- 0.0100\ M \times 125$ mL $= 0.75$ mmol
$C_{H4Y} = 0.75$ mmol/250 mL $= 3.0_0 \times 10^{-3}\ M$
$(5.00 \times 10^{-3})/\{[Pb^{2+}](3.0_0 \times 10^{-3})\} = 2.7_8 \times 10^7$
$[Pb^{2+}] = 6.0_0 \times 10^{-8}\ M$
$pPb = -\log 6.0_0 \times 10^{-8} = 7.22$

(b) $3.9_0 \times 10^{17}$

(1) $pPb = 1.60$ [as in (a)]

(2) $pPb = 2.12$ [as in (a)]

(3) From (a): $[PbY^{2-}] = 7.14 \times 10^{-3}\ M$
$(7.14 \times 10^{-3})/(x^2) \approx 3.9_0 \times 10^{17};\ x = 1.35 \times 10^{-10} = [Pb^{2+}]$
$pPb = -\log 1.35 \times 10^{-10} = 9.87$

(4) From (a): $C_{H4Y} = 3.0_0 \times 10^{-3}\ M$, $[PbY^{2-}] = 5.00 \times 10^{-3}\ M$
$(5.00 \times 10^{-3})/\{[PbY^{2-}](3.0_0 \times 10^{-3})\} = 3.9_0 \times 10^{17}$;
$[Pb^{2+}] = 4.2_7 \times 10^{-18}\ M$
$pPb = -\log 4.2_7 \times 10^{-18} = 17.37$

11. From Problem 9, $\alpha_4 = 2.5_3 \times 10^{-11}$
$\therefore\ K_f' = K_f \alpha_4 = 5.0 \times 10^{10} \times 2.5_3 \times 10^{-11} = 1.2_6$

Hence, at pH 3, the Ca-EDTA chelate is very weak. In view of the large differences in the K_f' values for calcium and lead, it should be possible to titrate lead in the presence of calcium.

12. $K_f' = \alpha_{Ni}\alpha_{Y4}K_f$
From Example 9.3, α_{Y4-} at pH 10 is 0.36. From Appendix Table C.4, K_f is 4.16×10^{18}. See Example 9.7 for the approach for solving the problem.

From Equation (9.19):
$\alpha_{Ni} = (\beta_0 + \beta_1[NH_3] + \beta_2[NH_3]^2 + \beta_3[NH_3]^3 + \beta_4[NH_3]^4 + \beta_5[NH_3]^5 + \beta_6[NH_3]^6)^{-1}$
$= (1 + 10^{2.67} \times 10^{-1} + 10^{4.79} \times 10^{-2} + 10^{6.40} \times 10^{-3} + 10^{7.47} \times 10^{-4} + 10^{8.10} \times 10^{-5} + 10^{8.06}$
$\times 10^{-6})^{-1}$
$= (7,467)^{-1} = 1.34 \times 10^{-4}$
$K_f' = 1.34 \times 10^{-4} \times 0.36 \times 4.16 \times 10^{18} = 2.01 \times 10^{14}$

13. $0.0500\ M \times 500.0$ mL $= (x$ mg $Na_2H_2Y \cdot 2H_2O)/(372.23$ mg/mmol$)$
$x = 9306$ mg $= 9.306$ g $Na_2H_2Y \cdot 2H_2O$

14. $M_{EDTA} \times 38.26$ mL $= (398.2$ mg$)/(100.09$ mg/mmol$)$
$M_{EDTA} = 0.1039_8$

15. Titer = mg $CaCO_3$/mL EDTA. Since the reaction is 1:1, mmol $CaCO_3$ = mmol EDTA.
    mmol $CaCO_3$ = M_{EDTA} x mL_{EDTA}
    (mg $CaCO_3$)/[100.1 (mg/mmol)] = 0.1000 M x 1.000 mL = 0.1000 mmol
    mg $CaCO_3$ = 0.1000 x 100.1 = 10.01 mg/mL EDTA

16. Water hardness = mg $CaCO_3$/L H_2O. The milligrams $CaCO_3$ in 100.0 mL
    (1 dL) = $1/10^{th}$ the water hardness.
    From Problem 13, 0.100 M EDTA is equivalent to 10.01 mg $CaCO_3$/mL, so 0.0100 M EDTA
    is equivalent to 1.001 mg $CaCO_3$/mL.
    mL_{EDTA} x 1.001 (mg $CaCO_3$/mL) x 10 (dL/L) = mg $CaCO_3$/L H_2O = water hardness.
    Water hardness/mL EDTA = 10.01 (mg $CaCO_3$/L H_2O)/mL EDTA

17. M_{Zn} = [632 mg/65.4 (mg/mmol)]/(1000 mL) = 0.00966 mmol/mL
    M_{EDTA} = [10.0 mL_{Zn} x 0.00966 (mmol/mL)]/(10.8 mL_{EDTA})
          = 0.00894 mmol/Ml
    mg_{Ca} = 0.00894 mmol/mL EDTA x 12.1 mL EDTA x 40.1 mg Ca/mmol
          = 4.34 mg
    ppm_{Ca} = mg/kg = 4.34 mg/1.50 x 10^{-3} kg = 2.89 x 10^3 ppm

18. [0.00122 M x 0.203 mL x 40.1 mg/mmol]/10^{-3} (dL/100 μL)
    = 9.93 mg Ca/dL serum 9.93 mg/dL = 99.3 mg/L
    (99.3 mg/L)/(40.1/2 mg/meq) = 4.95 meq/L

19. % KCN = {[0.1025 M x 34.95 mL x 2 (mmol CN^-/mmol Ag^+) x 65.12 mg/mmol]/(472.3 mg)}
    x 100% = 98.79%

20. EDTA titer = 2.69 mg $CaCO_3$/mL
        M_{EDTA} = [2.69 mg $CaCO_3$/100.1 (mg $CaCO_3$/mmol)]/(1 mL EDTA)
            = 0.0269 mmol/mL
    mg_{Cu} titrated = 0.0269 mmol/mL EDTA x 2.67 mL EDTA x 63.5 mg
    Cu/mmol = 4.56 mg
    ppm_{Cu} = mg/L = [4.56 (mg/50 mL) x 2]/(3 L) = 3.04 mg/L

21. $M_{Hg(NO3)2}$ x 1.12 mL = 0.0108 M x 2.00 mL x ½(mmol $Hg(NO_3)_2$/mmol NaCl)
    $M_{Hg(NO3)2}$ = 0.00964 M

    One-half mL serum was diluted to 5.00 mL and 2.00 mL (40.0%) of this was taken for
    titration, or 0.200 mL serum.
    mg Cl^-/L = [0.00964 M x 1.23 mL x 2 (mmol Cl^-/mmol $Hg(NO_3)_2$) x mg/mmol x 10^3 mL/L]/
    (0.200 mL serum) = 119 mg/L serum

22. PFP
    The formula weight of ZnO is 81.37 g/mol
    0.0100 M × 0.02552 L × 81.37 g/mol = 0.0208 g
    (0.0208 g/0.1021 g) × 100 = 20.4%

23. See text website for spreadsheet and graph of log K_f' values.

24. See text website for spreadsheet and graph of titration curve.

25. See text website for solution.

CHAPTER 10   GRAVIMETRIC ANALYSIS AND PRECIPITATION EQUILIBRIA

1. The solution is adjusted for optimum precipitating conditions, and the analyte is precipitated, digested to obtain a pure and filterable precipitate, filtered, washed to remove impurities (with a volatile electrolyte to prevent peptization), dried or ignited to a weighable form, and weighed in order to calculate the quantity of analyte.

2. The relative supersaturation $(Q - S)/S$. Q = concentration of mixed reagents before precipitation, S = solubility of precipitate.

3. Most favorable precipitates are obtained with minimal supersaturation, and the von Weimarn ratio predicts that favorable precipitation conditions are obtained by precipitating from dilute solution, slowly, with stirring (low Q), and from hot solution (high S).

4. The process of allowing a precipitate to stand in the presence of the mother liquor, often at elevated temperature, in order to obtain purer and larger crystals. Surface impurities and small crystals dissolve and the latter reprecipitate on the larger crystals, resulting in more perfect and larger crystals.

5. See answer to Question 3, and precipitate at as low a pH as possible to maintain quantitative precipitation.

6. Coprecipitation is the carrying down with the precipitate of normally soluble constituents in the solution. In occlusion, foreign ions are trapped in the crystal as it grows. These are difficult to get rid of, and reprecipitation may be required. Impurities adsorbed on the surface of the precipitate can often be removed by digestion and/or washing. Post precipitation is the slow precipitation (after a period of time) of a normally insoluble precipitate. It can be minimized by filtering as soon as possible. Isomorphous replacement is the formation of mixed crystals of two salts that are chemically similar. It is difficult to eliminate.

7. In order to remove the mother liquor and nonvolatile surface impurities and replace them by a volatile electrolyte.

8. In order to prevent peptization (formation of colloidal particles) of the precipitate. The electrolyte must be volatile at the temperature of drying or ignition and must not dissolve the precipitate.

9. Organic precipitation agents produce precipitates with very low solubility in water that have very favorable gravimetric factors since the organic reagents have high molecular weights. Depending on the reagent, selectivity can be high. Solubility can be adjusted by pH control.

10. g Na = g Na_2SO_4 x $(2Na/Na_2SO_4)$ = 50.0 x $[(2(22.99)/142.0]$ = 16.2 g

11. g BaSO$_4$ = g Na$_2$SO$_4$ x (BaSO$_4$/Na$_2$SO$_4$) = 50.0 x (233.4/142.0) = 82.2 g

12 (As$_2$O$_3$)/(2Ag$_3$AsO$_4$) = 197.8)/[2(462.5)] = 0.2138
    (2FeSO$_4$)/(Fe$_2$O$_3$) = [2(151.9)]/159.7 = 1.902
    (K$_2$O)/[2KB(C$_6$H$_5$)$_4$] = (94.20)/[2(358.3)] = 0.1314
    (3SiO$_2$)/(KAlSi$_3$O$_8$) = [3(60.08)]/278.4 = 0.6474

13 g CuO = 1.00 x [(4CuO)/Cu$_3$(A$_s$O$_3$)$_2$.2As$_2$O$_3$.Cu(C$_2$H$_3$O$_2$)$_2$]
           = 1.00 x [4(79.54)]/1.013 = 0.314 g
    g A$_s$O$_3$ = 1.00 x [3(197.8)]/1,013 = 0.586 g

14. % KBr = ([mg AgBr x (KBr)/(AgBr)]/mg sample x 100%
          = ([814.5 x (119.01)/(187.78)]/523.1) x 100% = 98.68%

15. g Fe = 0.4823 x 0.9989 = 0.4818 g
    g Fe$_2$O$_3$ = g Fe x (Fe$_2$O$_3$)/(2Fe) = 0.4818 g x (159.69)/(2 x 55.847) = 0.6888 g

16. % Al = (g Al(C$_9$H$_6$ON)$_3$ x [Al/Al(C$_9$H$_6$ON)$_3$]/g sample x 100%
         = [0.1862 g (26.98/459.5)/1.021 g] x 100% = 1.071%

17. Calculate the amount of iron required to give 100.0 mg Fe$_2$O$_3$.
    mg Fe = mg Fe$_2$O$_3$ x (2Fe/Fe$_2$O$_3$) = 100.0 mg x (2 x 55.85/159.7) = 69.94 mg

    The minimum iron content is 11%, so
    69.94 mg x (100/11) = 636 mg minimum sample needed.
    18. g MgCl$_2$ = 0.12 g x 0.95 = 0.11$_4$ g
    g Cl$^-$ = 0.11$_4$ x (2Cl/MgCl$_2$) = 0.11$_4$ x (2 x 35.4/95.2) = 0.085 g
    mmol Cl$^-$ = mmol Ag$^+$ = (85 mg)/(35 mg/mmol) = 2.4 mmol

    Total mmol Ag$^+$ to be added = 2.4 + 0.2 = 2.6 mmol
    x = 26 mL AgNO$_3$

19 2NH$_3$ → (NH$_4$)$_2$PtCl$_6$ → Pt
    ∴ % NH$_3$ = [g Pt x (2 NH$_3$/Pt)/g sample] x 100%
      = [0.100 x (x 17.03)/195.1)/1.00] x 100% = 1.75%

20. % Cl = [g AgCl x (Cl/AgCl)/x g] x 100%
    Since % Cl = g AgCl, these cancel. Thus,
    x = (35.45/143.3) x 100 = 24.74 g

21. Assume 1.000 g of BaSO$_4$, so the % FeS$_2$ is 10.00.
    [1.000 x (FeS$_2$/2 BaSO$_4$)/x g] x 100% = 10.00%
    [1.000 x (120.0)/2(233.4)/x g] x 100% = 10.00%
    x = 2.571 g of ore

22. $x = $ g BaO $y = $ g CaO

    (1) $x + y = 2.00$ g

    (2) x (BaSO$_4$/BaO) + y (CaSO$_4$/CaO) = 4.00 g
        x (233.4/153.3) + y (136.1/56.08) = 4.00 g
        $1.52x + 2.43y = 4.00$g

    Solution of the simultaneous equations gives:
    $x = 0.95$ g BaO
    %Ba = [0.95 x (137.3/153.3)/2.00] x 100% = 42.5%
    %Ca = [1.05 x (40.08/56.08)/2.00] x 100% = 37.5%

23. Assume 100.0 g sample. X = g CaSO$_4$
    CaSO$_4$ + g BaSO$_4$ = 100.0
        $x + x$(Ca/CaSO$_4$)(1/2)(BaSO$_4$/Ba) = 100.0

$$\frac{\text{gCa}}{\text{g Ba}}$$

    $x + x$ (40.08/136.1)(1/2)(233.4/137.3) = 100.0

    $x = 79.98$

    % CaSO$_4$ = (79.98/100.0) x 100% = 79.98%

24. There are two unknowns, and so there must be two equations.
    $x = $ g AgCl $y = $ g AgBr

    (1) $x + y = 2.00$

    (2) g Ag from AgCl + g Ag from AgBr = 1.300
        x (Ag/AgCl) + y (Ag/AgBr) = 1.300g
        x (107.9/143.3) + y (107.9/187.8) = 1.300g
        $0.7530x + 0.5745y = 1.300$g

    Solution of the simultaneous equations gives:
    $x = 0.846$ g AgCl
    $y = 1.154$ g AgBr

25. (a) $K_{sp} = [Ag^+][SCN^-]$

    (b) $K_{sp} = [La^{3+}][IO_3^-]^3$

    (c) $K_{sp} = [[Hg_2^{2+}][Br^-]^2$

(d) $K_{sp} = [Ag^+][[Ag(CN)_2^-]$

(e) $K_{sp} = [Zn^{2+}]^2[Fe(CN)_6^{4-}]$

(f) $K_{sp} = [Bi^{3+}]^2[S^{2-}]^3$

26. $(7.76 \times 10^{-3} \text{ g/L})/(590 \text{ g/mol}) = 1.32 \times 10^{-5} \text{ } M$
$BiI_3 = Bi^{3+} + 3I^-$
$[Bi^{3+}] = 1.32 \times 10^{-5} \text{ } M$
$[I^-] = 3 \times 1.32 \times 10^{-5} \text{ } M = 3.96 \times 10^{-5} \text{ } M$
$K_{sp} = [Bi^{3+}][I^-]^3 = (1.32 \times 10^{-5})(3.96 \times 10^{-5})^3 = 8.20 \times 10^{-19}$

27. $Ag_2CrO_4 = 2Ag^+ + CrO_4^{2-}$
$\qquad\qquad 2s \qquad\quad s$
$[Ag^+]^2[CrO_4^{2-}] = 1.1 \times 10^{-12}$
$(2s)^2(s) = 1.1 \times 10^{-12}$
$s = 6.5 \times 10^{-5}$
$[Ag^+] = 2 \times 6.5 \times 10^{-5} = 1.3 \times 10^{-4} \text{ } M$
$[CrO_4^{2-}] = 6.5 \times 10^{-5} \text{ } M$

28. mmol $Ba^{2+} = 0.100 \times 25.0 = 2.50$ mmol
mmol $CrO_4^{2-} = 0.200 \times 15.0 = 3.00$ mmol
excess $CrO_4^{2-} = 3.00 - 2.50 = 0.50$ mmol/40 mL $= 0.012_5 \text{ } M$
$BaCrO_4 = Ba^{2+} + CrO_4^{2-}$
$\qquad\qquad\quad s \qquad (s + 0.0125)$
$[Ba^{2+}][CrO_4^2] = 2.4 \times 10^{-10}$
$[Ba^{2+}] = (2.4 \times 10^{-10})/(0.012_5) = 1.9 \times 10^{-8} \text{ } M$

29. $[Ag^+]^3[PO_4^{3-}] = 1.3 \times 10^{-20}$
$[PO_4^{3-}] = (1.3 \times 10^{-20})/(0.10)^3 = 1.3 \times 10^{-17} \text{ } M$

30. For PO_4^{3-}:
$[Ag^+]^3[PO_4^{3-}] = 1.3 \times 10^{-20}$
$[Ag^+] = \sqrt[3]{(1.3x10^{-20}/(0.10)} = 5.1 \times 10^{-7} \text{ } M$

For Cl^-:
$[Ag^+][Cl^-] = 1.0 \times 10^{-10}$
$[Ag^+] = (1.0 \times 10^{-10})/(0.10) = 1.0 \times 10^{-9} \text{ } M$

31. $Al(OH)_3 = Al^{3+} + 3OH^-$
$[Al^{3+}][OH^-]^3 = 2 \times 10^{-32}$
$(0.10)[OH^-]^3 = 2 \times 10^{-32}$
$[OH^-] = \sqrt[3]{2x10^{-32}/0.10} = 6 \times 10^{-11} \text{ } M$
$pOH = -\log 6 \times 10^{-11} = 10.2$
$pH = 14.0 - 10.2 = 3.8$

32. $Ag_3AsO_4 = 3Ag^+ + AsO_4^{3-}$

    $3s \qquad s$

    $[Ag^+]^3[AsO_4^{3-}] = 1.0 \times 10^{-22}$

    $(3s)^3(s) = 1.0 \times 10^{-22}$

    $s = 1.4 \times 10^{-6}$

    $[Ag_3AsO_4] = 1.4 \times 10^{-6} \ M$

    $1.4 \times 10^{-6} \ mol/L \times 0.25 \ L \times 463 \ g/mol = 1.6 \times 10^{-4} \ g \ Ag_3AsO_4$

33. Let s = solubility of Ag_2CrO_4

    $Ag_2CrO_4 \ = \ 2Ag^+ \ + \ CrO_4^{2-}$

    $2s \qquad (s + 0.10)$

    $[Ag^+][CrO_4^{2-}] = 1.1 \times 10^{-12}$

    $(2s)(0.10 + s) = 1.1 \times 10^{-12}$

    Assume s is negligible compared to 0.10. Then,

    $0.40 \ s^2 = 1.1 \times 10^{-12}$

    $s = 1.7 \times 10^{-6} \ M$

34. $AB = A^{2+} + B^{2-}$ $\qquad\qquad\qquad\qquad$ $AC_2 = A^{2+} + 2C^-$

    $s \quad\ \ s$ $\qquad\qquad\qquad\qquad\quad$ $s \qquad 2s$

    $[A^{2+}][B^{2-}] = 4 \times 10^{-18}$ $\qquad\quad$ $[A^{2+}][2C^-]^2 = 4 \times 10^{-18}$

    $(s)(s) = 4 \times 10^{-18}$ $\qquad\qquad\quad\ $ $(s)(2s)^2 = 4 \times 10^{-18}$

    $s = 2 \times 10^{-9} \ M$ $\qquad\qquad\qquad\ $ $s = 1 \times 10^{-6} \ M$

    $\therefore$ Compound AC_2 is more soluble.

35. $Bi_2S_3 = 2Bi^{3+} + 3S^{2-}$

    Let s be the solubility. Then $[Bi^{3+}] = 2s$, $[S^{2-}] = 3s$.

    $[Bi^{3+}]^2[S^{2-}]^3 = 1 \times 10^{-97}$

    $(2s)^2(3s)^3 = 1 \times 10^{-97}$

    $s = 1._6 \times 10^{-20} \ M$

    $HgS = Hg^{2+} + S^{2-}$

    Let s = the solubility. Then $[Hg^{2+}] = [S^{2-}] = s$

    $[Hg^{2+}][S^{2-}] = 4 \times 10^{-53}$

    $(s)(s) = 4 \times 10^{-53}$

    $s = 6 \times 10^{-27} \ M$

    Bi_2S_3 is 4×10^7 times more soluble than HgS!

36. No more than 0.1% of the Ba^{2+} must remain, or <0.20mg (in 100 mL).

    After precipitation,

    $[Ba^{2+}] = (2.0 \times 10^{-3} \ mg/mL)/(137 \ mg/mmol) = 1.5 \times 10^{-5} \ M$

    $[Ba^{2+}][F^-] = 1.7 \times 10^{-6}$

    $(1.5 \times 10^{-5})[F^-] = 1.7 \times 10^{-6}$

    $[F^-] = 0.33 \ M$

    This concentration of excess fluoride is attainable, so the analysis in principal would work.

37. (a) $K_{sp} = a_{Ba2+} \cdot a_{SO42-} = [Ba^{2+}]f_{Ba2+}[SO_4^{2-}]f_{SO42-} = K_{sp} \ f_{Ba2+} \cdot f_{SO42-}$

    (b) $K_{sp} = a_{Ag+}^2 \cdot a_{CrOr42-} = [Ag^+]^2 f_{Ag+}^2[CrO_4^{2-}]f_{CrO42-} = K_{sp} \ f_{Ag+}^2 \cdot f_{CrO42-}$

38. $\mu = 0.0375$. From calculations using Equation 4.19, $f_{Ba^{2+}} = 0.502$, $f_{SO_4^{2-}} = 0.485$.

From the appendix, K_{sp} at zero ionic strength (K_{sp}^0) is 1.0×10^{-10}.
$K_{sp} = [Ba^{2+}][SO_4^{2-}]f_{Ba^{2+}} f_{SO_4^{2-}}$
$1.0 \times 10^{-10} = (s)(s)(0.502)(0.495)$
$s = 2.0 \times 10^{-5}$ M

39. $CaF_2 \quad = \quad Ca^{2+} \quad + \quad 2F^-$
$\qquad\qquad 0.015 + 1/2x \qquad x$
$\qquad\qquad \approx 0.015$

$K_{sp} = [Ca^{2+}]f_{Ca^{2+}} \cdot [F^-]^2 f_F^{-2} = K_{sp}f_{Ca^{2+}}f_F^{-2}$
$K_{sp} = K_{sp}^0/f_{Ca^{2+}}f_F^{-2} = [Ca^{2+}][F^-]^2$

From the appendix, $K_{sp}^0 = 4.0 \times 10^{-11}$
The solution contains 0.015 M $Ca(NO_3)_2 + 0.025$ M $NaNO_3$ (neglect amount of F_2 in solution).
$\mu = \frac{1}{2}([Ca^{2+}](2)^2 + [Na^+](1)^2 + [NO_3^-](1)^2)$
$= \frac{1}{2}[(0.015)(4) + (0.025)(1) + (0.040)(1)] = 0.062$

From Reference 9 in Chapter 6,
$\alpha_{Ca^{2+}} = 6 \quad \alpha_{F^-} = 3.5$

From Equation 6.19 in Chapter 6,
$f_{Ca^{2+}} = 0.46$; $f_F^- = 0.76$
$K_{sp} = (4.0 \times 10^{-11})/(0.46)(0.76) = 1.1_4 \times 10^{-10}$
$(0.015)(x)^2 = 1.1_4 \times 10^{-10}$
$x = 8.7 \times 10^{-5}$ $M = [F^-]$
$(8.7 \times 10^{-5}$ mol/L$)(0.025$ L$)(19.0$ g/mol$) = 4.1 \times 10^{-4}$ g F^- in solution.

40. See the text website for the spreadsheet. The calculated value is 16.298% P_2O_5. For a 0.5267 g sample and 2.0267 g precipitate, the % P_2O_5 is 14.55_4 %.

41. See the text website for the spreadsheet and graph.

42. See the text website for the spreadsheet and graph.

43. See the text website for the spreadsheet calculation.

CHAPTER 11 PRECIPITATION REACTIONS AND TITRATIONS

1. The Volhard titration involves adding excess standard silver nitrate to a chloride solution and then back titrating the excess silver with standard potassium thiocyanate solution. The AgCl precipitate is first removed by filtration, or else nitrobenzene is added to prevent its reaction with SCN^-.The indicator is Fe^{3+} and the end point is marked by formation of the red $Fe(SCN)^{2+}$ complex. The Fajans titration involves the direct titration of chloride. The end point is marked by adsorption of the indicator, dichlorofluoroscein, which causes it to turn pink. Dextrin is added to minimize coagulation at the end point. The Volhard method must be used for strong acid solution, because the adsorption indicator in the Fajan titration is too slightly dissociated in acid solution for the anion form to be adsorbed.

2. The ionized form of the indicator, which has a charge opposite that of the titrating ion, becomes adsorbed on the surface of the precipitate surface, which attracts the indicator ion. The adsorbed indicator has a color distinctly different from the nonadsorbed indicator, possibly due to complexation with the titrant in the precipitate.

3   From the appendix, K_{sp} of $AgIO_3 = 3.1 \times 10^{-8}$ and K_a of $HIO_3 = 2 \times 10^{-1}$. The equilibiria are
$AgIO_3 = Ag^+ + IO_3^-$
$IO_3^- + H^+ = HIO_3$
$K_{sp} = [Ag^+][IO_3^-] = [Ag^+]C_{HIO3} \alpha_1$

Where α_1 is the fraction of the total iodate (C_{HIO3}) that exists as IO_3^-. Then,
$K_{sp}/\alpha_1 = K_{sp}' = [Ag^+]C_{HIO3} = s^2$

where s is the water solubility. First calculate α_1.
$C_{HIO3} = [HIO_3] + [IO_3^-]$

But, $[HIO_3] = [H^+][IO_3^-]/K_a$. So,
$C_{HIO3} = [H^+][IO_3^-]/K_a + [IO_3^-]$

from which
$1/\alpha_1 = C_{HIO3}/[IO_3^-] = [H^+]/K_a + 1 = 1.5$ (at 0.100 M acid)
$K_{sp}' = (3.1 \times 10^{-8})(1.5) = s^2$
$s = 2.1 \times 10^{-4} M = [Ag^+] = C_{HIO3}$

(This is only increased from $1.8 \times 10^{-4} M$ in the absence of acid, since HIO_3 is a fairly strong acid.)
$\alpha_1 = 0.67; \ \alpha_0 = 1 - 0.67 = 0.33$
$\therefore \ [IO_3^-] = (2.1 \times 10^{-4})(0.67) = 1.5 \times 10^{-4} M$
$[HIO_3] = (2.1 \times 10^{-4})(0.33) = 6.9 \times 10^{-5} M$

4. From the appendix, K_{sp} of $CaF_2 = 4.0 \times 10^{-11}$ and K_a of $HF = 6.7 \times 10^{-4}$. The equilibria are
$$CaF_2 = Ca^{2+} + 2F^-$$
$$F^- + H^+ = HF$$
$$K_{sp} = [Ca^{2+}][F^-]^2$$
(since $[F^-] = C_{HF} \alpha_1$).
$$K_{sp}/\alpha_1^2 = K_{sp}' = [Ca^{2+}]C_{HF}^2 = (s)(2s)^2$$

As in the preceding problem,
$$1/\alpha_1 = [H^+]/K_a + 1 = 1.5_1 \times 10^2 \text{ (at 0.100 } M \text{ acid)}$$
$$K_{sp}' = (4.0 \times 10^{-11})(1.5_1 \times 10^2)^2 = (s)(2s)^2$$
$$s = 6.1 \times 10^{-3} M = [Ca^{2+}] = \frac{1}{2} C_{HF}$$
$$\alpha_1 = 6.6_2 \times 10^{-3}; \quad \alpha_0 = 1 - 6.6_2 \times 10^{-3} = 0.993$$
Since $C_{HF} = 2 \times 6.1 \times 10^{-3} = 1.2_1 \times 10^{-2} M$,
$$[HF] = (1.2_1 \times 10^{-2})(0.993) = 1.2_0 \times 10^{-2} M$$
$$[F^-] = (1.2_1 \times 10^{-2})(6.6_2 \times 10^{-3}) = 8.0_1 \times 10^{-5} M$$

Essentially all of the fluoride exists as HF and its concentration is twice that of the calcium. About 12% of the HCl was consumed in forming HF. Recalculation (reiteration) using 0.088 M HCl would result in a minor correction in the calculation ($s = 5.6 \times 10^{-3} M$).

5. From the appendix, K_{sp} of $PbS = 8 \times 10^{-28}$ and $K_{a1} = 9.1 \times 10^{-8}$, $K_{a2} = 1.2 \times 10^{-15}$ for H_2S. The equilibria are:
$$PbS = Pb^{2+} + S^{2-} \tag{1}$$
$$S^{2-} + H^+ = HS^- \tag{2}$$
$$HS^- + H^+ = H_2S \tag{3}$$
$$K_{sp} = [Pb^{2+}][S^{2-}] = [Pb^{2+}]C_{H2S}\alpha_2 \tag{4}$$
$$K_{sp}/\alpha_2 = K_{sp}' = [Pb^{2+}]C_{H2S} = s^2 \tag{5}$$

We calculate for the diprotic acid that
$$\alpha_0 = ([H^+]^2)/([H^+]^2 + K_{a1}[H^+] + K_{a1}K_{a2}) = 1.00 \text{ (for 0.0100 } M \text{ acid)}$$
$$\alpha_1 = (K_{a1}[H^+])/([H^+]^2 + K_{a1}[H^+] + K_{a1}K_{a2}) = 9.1 \times 10^{-6}$$
$$\alpha_2 = (K_{a1}K_{a2})/([H^+]^2 + K_{a1}[H^+] + K_{a1}K_{a2}) = 1.0_9 \times 10^{-18}$$
$$K_{sp}' = (8 \times 10^{-28})/(1.0_9 \times 10^{-18}) = s^2$$
$$s = 2.7 \times 10^{-5} M = [Pb^{2+}] = C_{H2S}$$
$$[H_2S] = C_{H2S}\alpha_0 = (2.7 \times 10^{-5})(1.00) = 2.7 \times 10^{-5} M$$
$$[HS^-] = C_{H2S}\alpha_1 = (2.7 \times 10^{-5})(9.1 \times 10^{-6}) = 2.7 \times 10^{-10} M$$
$$[S^{2-}] = C_{H2S}\alpha_2 = (2.7 \times 10^{-5})(1.0_9 \times 10^{-18}) = 2.9 \times 10^{-18} M$$
Virtually all the sulfide is in the H_2S form.

6. From the appendix, K_{sp} of AgCl is 1.0×10^{-10}. the equilibria are:
$$AgCl = Ag^+ + Cl^- \tag{1}$$
$$Ag^+ + en = Ag(en)^+ \tag{2}$$
$$Ag(en)^+ + en = Ag(en)_2^+ \tag{3}$$
$$K_{sp} = [Ag^+][Cl^-] = Ag_T\alpha_M[Cl^-] \tag{4}$$
$$K_{sp}/\alpha_M = K_{sp}' = Ag_T[Cl^-] = s^2 \tag{5}$$

For a complex with two ligands [see Eqaution 9.20 and Example 9.5), we calculate that

$\alpha_{Ag}= 1/(K_{f1}K_{f2}[en]^2 + K_{f1}[en] + 1) = 1.4_3 \times 10^{-6}$ (for 0.100 M en).

$\beta_1 = (K_{f1}[en])/(K_{f1}K_{f2}[en]^2 + K_{f1}[en] + 1) = 7.1 \times 10^{-3}$

$\beta_2 = (K_{f1}K_{f2}[en]^2)/(K_{f1}K_{f2}[en]^2 + K_{f1}[en] + 1) = 1.00$

$K_{sp}' = (1.0 \times 10^{-10})/(1.4_3 \times 10^{-6}) = s^2$

$s = 8.4 \times 10^{-3} M = Ag_T = [en]$

$[Ag^+] = Ag_T\alpha_{Ag} = (8.4 \times 10^{-3})(1.4_3 \times 10^{-6}) = 1.2_0 \times 10^{-8} M$

$[Ag(en)^+] = Ag_T\beta_1 = (8.4 \times 10^{-3})(7.1 \times 10^{-3}) = 5.9_6 \times 10^{-5} M$

$[Ag(en)_2^+] = Ag_T\beta_2 = (8.4 \times 10^{-3})(1.00) = 8.4 \times 10^{-3} M$

7. The Equilibria are:

$AgIO_3 = Ag^+ + IO_3^-$

$IO_3^- + H^+ = HIO_3$

$H_2O = H^+ + OH^-$

$HNO_3 = H^+ + NO_3^-$

The equilibrium constant expressions are:

$K_{sp} = [Ag^+][IO_3^-] = 3.1 \times 10^{-8}$ (1)

$K_a = [H^+][IO_3^-]/[HIO_3] = 2 \times 10^{-1}$ (2)

$K_w = [H^+][OH^-] = 1.00 \times 10^{-14}$ (3)

The mass balance expressions are:

$[Ag^+] = [IO_3^-] + [HIO_3] = C_{HIO3}$ (4)

$[H^+] = [NO_3^-] + [OH^-] - [HIO_3]$ (5)

$[NO_3^-] = 0.100 M$ (6)

The charge balance expression is:

$[H^+] + [Ag^+] = [IO_3^-] + [NO_3^-] + [OH^-]$ (7)

There six unknowns ($[H^+]$, $[OH^-]$, $[NO_3^-]$, $[HIO_3]$, $[IO_3^-]$, $[Ag^+]$) and 6 independent equations.

Simplifying assumptions:

(1) Assume $[OH^-]$ is very small.

(2) Assume $[HIO_3]$ is small compared to $[H^+]$, since K_{sp} is small.

(3) K_a is large, but so is the acidity. Let's assume, though, that $[IO_3^-] \gg [HIO_3]$.

i.e., $\alpha_1 \gg \alpha_2$ (it is actually 67% of the total – see Problem 3).

With these assumptions, from (5) and (6),

$[H^+] = 0.100 + [OH^-] - [HIO_3] \approx 0.100$ (8)

From (1)

$[Ag^+] = K_{sp}/[IO_3^-]$ (9)

From assumption (3), $[Ag^+] \approx [IO_3^-]$

$[Ag^+] = K_{sp}/[Ag^+] = 3.1 \times 10^{-8}/[Ag^+]$ (10)

$[Ag^+] = 1.8 \times 10^{-4} \ M$

This compares with $2.1 \times 10^{-4} \ M$ calculated in Problem 3, being 14% lower because of the assumption that HIO_3 does not form (which increases solubility). To be more correct, exact solution of the simultaneous equations is needed, or we can calculate α_1 (it is 0.67, so $[HIO_3] = 0.33[Ag^+]$ and following from (9) and (2),

$[Ag^+] = K_{sp}[H^+]/K_a[HIO_3] = K_{sp}[H^+]/K_a 0.33[Ag^+]$

$= (3.1 \times 10^{-8})(0.100)/(2 \times 10^{-1})(0.33)[Ag^+]$

$[Ag^+] = 2.1 \times 10^{-4} \ M$

The same is obtained by substituting $0.67[Ag^+]$ for IO_3^- in (9).

8. Equilibria:

$PbS = Pb^{2+} + S^{2-}$

$S^{2-} + H^+ = HS^-$

$HS^- + H^+ = H_2S$

Equilibrium expressions:

$[Pb^{2+}][S^{2-}] = K_{sp} = 8 \times 10^{-28}$ (1)

$[H^+][HS^-]/[H_2S] = K_{a1} = 9.1 \times 10^{-8}$ (2)

$[H^+][S^{2-}]/[HS^-] = K_{a2} = 1.2 \times 10^{-15}$ (3)

Mass balance expressions:

$[Pb^{2+}] = [S^{2-}] + [HS^-]$ (4)

$[H^+] = [Cl^-] + [OH^-] - [HS^-] - [H_2S]$ (5)

$[Cl^-] = 0.0100 \ M$ (6)

6 unknowns ($[Pb^{2+}], [S^{2-}], [HS^-], [H_2S], [H^+], [OH^-]$) and 6 independent equations

Assume: $[H^+] \gg [OH^-], [HS^-]$ and $[H_2S]$, and $[H_2S] \gg [HS^-]$ and $[S^{2-}]$

Then from (4),

$[Pb^{2+}] \approx [H_2S]$

From (5),

$[H^+] \approx 0.0100 \ M$

Calculate $[Pb^{2+}]$:

From (1), $[Pb^{2+}] = K_{sp}/[S^{2-}]$ (7)

From (3), $[S^{2-}] = K_{a2}[HS^-]/[H^+]$ (8)

From (2), $[HS^-] = K_{a1}[H_2S]/[H^+] \approx K_{a1}[Pb^{2+}]/[H^+]$ (9)

$\therefore \ [S^{2-}] = K_{a1}K_{a2}[Pb^{2+}]/[H^+]^2$ (10)

Substituting (10) in (7):

$[Pb^{2+}] = K_{sp}[H^+]^2/K_{a1}K_{a2}[Pb^{2+}]$

$= [(8 \times 10^{-28})(1.00 \times 10^{-2})^2]/\{(9.1 \times 10^{-8})(1.2 \times 10^{-15})[Pb^{2+}]\}$

$= 7.3 \times 10^{-10}/[Pb^{2+}]$

Solving, $[Pb^{2+}] = 2._7 \times 10^{-5}\ M$

This is the same answer as calculated in Problem 5.

9. Equilibria

$Ag^+ + en = Ag(en)^+$ $K_{f1} = [Ag(en)^+]/[Ag^+][en] = 5.0 \times 10^4$ (1)

$Ag(en)^+ + en = Ag(en)_2^+$ $K_{f2} = [Ag(en)_2^+]/[Ag(en)^+][en] = 1.4 \times 10^3$ (2)

$AgCl = Ag^+ + Cl^-$ $K_{sp} = [Ag^+][Cl^-] = 1.0 \times 10^{-10}$ (3)

Mass balance expressions:

$[en] = 0.100\ M - [Ag(en)^+] - [Ag(en)_2^+]$ (4)

$[Ag^+] = [Cl^-] - [Ag(en)^+] - [Ag(en)_2^+]$ (5)

or

$[Cl^-] = [Ag^+] = [Ag(en)^+] - [Ag(en)_2^+]$

Charge balance expression:

$[Ag^+] + [Ag(en)^+] - [Ag(en)_2^+] = [Cl^-]$ (6) ≡ (5)

5 unknowns: $[Ag^+]$, $[Ag(en)^+]$, $[Ag(en)_2^+]$, $[en]$, $[Cl^-]$
5 independent equations

Simplifying assumptions :

Since K_{f1} and K_{f2} are fairly large, assume $[Ag(en)_2^+] \gg [Ag(en)^+]$ and $[Ag^+]$

∴ $[Cl^-] \approx [Ag(en)_2^+]$ = solubility of AgCl = s

Assume solubility is small, and so $[en] \approx 0.100\ M$

From (3), $[Cl^-] = K_{sp}/[Ag^+] = s$ (7)

From (1), $[Ag^+] = [Ag(en)^+]/K_{f1}[en]$ (8)

From (2), $[Ag(en)^+] = [Ag(en)_2^+]/K_{f2}[en]$ (9)

Substituting (9) in (8),

$[Ag^+] = [Ag(en)_2^+]/K_{f1}K_{f2}[en]^2 \approx [Cl^-]/K_{f1}K_{f2}[en]^2$ (10)

∴ $[Cl^-] = K_{sp}K_{f1}K_{f2}[en]^2/[Cl^-]$

$[Cl^-] = \sqrt{K_{sp}K_{f1}K_{f2}[en]^2} = \sqrt{(1.0 x 10^{-10})(5.0 x 10^4)(1.4 x 10^3)(0.100)^2}$

$= 8.4 \times 10^{-3}\ M = s$

This is the same solubility calculated in Problem 6.

Note that in both Problem 6 and this one, we neglected the amount of en consumed ($= 2 \times 8.4 \times 10^{-3} = 0.016_8$ M, leaving 0.083_2 M). We could reiterate either problem using this new concentration of en, giving $s = 7.0 \times 10^{-3}$ M.

10. mg Cl in sample = $(0.1182$ $M \times 15.00$ mL $- 0.101$ $M \times 2.38$ mL$) \times 35.45$ mg/mmol
    = 54.34 mg
    $(54.34$ mg/10.00 mL$) \times 10^3$ mL/L $\times 10^{-3}$ g/mg = 5.434 g/L

11. In order for Ag_2CrO_4 to precipitate, the Ag^+ concentration is given by
    $[Ag^+]^2(0.0011) = 1.1 \times 10^{-12}$
    $[Ag^+] = 3.2 \times 10^{-5}$ M

    This must come from the solubility of AgCl plus excess titrant. Calculate the solubility of AgCl in the presence of 3.2×10^{-5} M Ag^+.

    $$AgCl \quad = \quad Ag^+ \quad + \quad Cl^-$$
    $$\qquad\qquad 3.2 \times 10^{-5} \qquad s$$
    $(3.2 \times 10^{-5})(s) = 1.0 \times 10^{-10}$
    $s = 3.1 \times 10^{-6}$ $M = [Ag^+]$ produced from the precipitate.

    Total mol Ag^+ = 3.2×10^{-5} $M \times 100$ mL = $\qquad$ 3.2×10^{-3}
    mmol Ag^+ from ppt. = 3.1×10^{-6} $M \times 100$ mL = $\underline{0.31 \times 10^{-3}}$
    mmol excess Ag^+ from titrant =
    2.0×10^{-3}
    0.100 $M \times x$ mL = 2.0×10^{-3} mmol
    $x = 0.029$ mL excess titrant

12. See text website for spreadsheet and plot of titration curve.

13. PFP
    Because K_{sp} ($Mn(OH)_2$) < K_{sp} ($Ca(OH)_2$), when adding NaOH to the mixture, $Mn(OH)_2$ will precipitate before $Ca(OH)_2$ does.

    Should 99.0% of Mn^{2+} be precipitated out of the solution, the concentration of Mn^{2+} left in the solution will be:

    $$\frac{0.10 \text{ mol/L} \times V \times 1\%}{V} = 1.0 \times 10^{-3} M$$

    $K_{sp} = [Mn^{2+}][OH^-]^2 = 1.6 \times 10^{-13}$, $[OH^-] = 1.2_6 \times 10^{-5}$, i.e., the concentration of OH^- in the solution when 99.0 % Mn^{2+} has been precipitated out.

    For $Ca(OH)_2 \leftrightarrow Ca^{2+} + 2OH^-$ in the solution,
    $Q = [Ca^{2+}][OH^-]^2 = 0.10 \times (1.2_6 \times 10^{-5})^2 = 1.6 \times 10^{-11} \ll K_{sp}$ ($Ca(OH)_2$). No $Ca(OH)_2$ precipitate will form. Therefore, it is possible to separate 99.0% of Mn^{2+} from Ca^{2+} without precipitation of $Ca(OH)_2$.

CHAPTER 12   ELECTROCHEMICAL CELLS AND ELECTRODE POTENTIALS

1. An oxidizing agent takes on electrons from a reducing agent. The former is reduced to a lower valence state and the latter is oxidized to a higher oxidation state.

2. The Nernst equation defines the potential of an electrode in a solution containing a redox couple: $aOx + ne^- = bRed$
$$E = E^0 - (2.303\ RT/nF)\log([Red]^b/[Ox]^a)$$

3. The standard potential (E^0) is the potential of an electrode in a solution relative to the normal hydrogen electrode, with all species at unit activity. The formal potential ($E^{0'}$) is the potential of an electrode under specified solution conditions.

4. A salt bridge prevents mixing of two solutions but allows charge transfer between them.

5. Normal hydrogen electrode (N.H.E.) or standard hydrogen electrode (S.H.E.). The standard potential of the half reaction $2H^+ + 2e^- = H_2$ is arbitrarily defined as zero and all other potentials are referred to this one.

6. A poor reducing agent.

7. About $0.2 - 0.3$ V.

8. They predict whether there is sufficient driving force (i.e., potential) for the reaction, but they say nothing of the rate of the reaction.

9. O_3, $HClO$, Hg^{2+}, H_2SeO_3, H_3AsO_4, Cu^{2+}, Co^{2+}, Zn^{2+}, K^+

10. Ni, H_2S, Sn^{2+}, V^{3+}, I^-, Ag, Cl^-, Co^{2+}, HF

11. (a) Fe^{2+} - MnO_4^-

    (b) Fe^{2+} - Ce^{4+} ($HClO_4$)

    (c) H_3AsO_3 - MnO_4^-

    (d) Fe^{3+} - Ti^{2+}

12. PFP

Oxidizing agent	Reducing agent
(a) VO^{2+}	Sn^{2+}
(b) $Fe(CN)_6^{3-}$	Fe^{2+}
(c) Ag^+	Cu
(d) I_2	I_2
(e) H_2O_2	Fe^{2+}

13. PFP

$$5Fe^{2+} + MnO_4^- + 8H^+ \Leftrightarrow 5Fe^{3+} + Mn^{2+} + 4H_2O$$

$$N_{MnO4-} \times V_{MnO4-} = N_{Fe2+} \times V_{Fe2+} = Weight_{Fe} \div Equivalent\ mass_{Fe}$$

The unit of normality (N) is equivalent per liter (equivalent/L)
The equivalent mass of Fe is 55.845 g/equivalent
$0.1023\ N \times 0.02875\ L = Weight_{Fe} \div 55.845\ g/equivalent$

$Weight_{Fe} = 0.1642\ g$
$(0.1642\ g/1.0512\ g) \times 100 = 15.62\%$

14. (a) $Pt/Fe^{2+}, Fe^{3+}//Cr_2O_7^{2-}, Cr^{3+}, H^+/Pt$

    (b) $Pt/I^-, I_2//IO_3^-, I_2, H^+/Pt$

    (c) $Zn/Zn^{2+}//Cu^{2+}/Cu$

    (d) $Pt/H_2SeO_3, SeO_4^{2-}, H^+//Cl_2, Cl^-/Pt$

15. (a) $2V^{2+} + PtCl_6^{2-} = 2V^{3+} + PtCl_4^{2-} + 2Cl^-$

    (b) $Ag + Fe^{3+} + Cl^- = \underline{AgCl} + Fe^{2+}$

    (c) $3Cd + ClO_3^- + 6H^+ = 3Cd^{2+} + Cl^- + 3H_2O$

    (d) $2I^- + H_2O_2 + 2H^+ = I_2 + H_2O$

16. $E = 1.52 - (0.059/5) \log [Br_2]^{1/2}/([BrO_3^-][H^+]^6)$
    $[H^+] = 10^{-2.5} = 10^{-5} \times 10^{-3} = 3.2 \times 10^{-3}\ M$
    $E = 1.52 - (0.059/5) \log [(0.20)^{1/2}/(0.50)(3.2 \times 10^{-3})^6] = 1.24V$

17. Since I^- is in large excess, I_3^- is formed instead of I_2:
    $H_2O_2 + 3I^- + 2H^+ = 2H_2O + I_3^-$

    Since both $[I^-]$ and $[I_3^-]$ are known, use this couple for calculations,

    mmol I_3^- formed = mmol H_2O_2 at start = 0.10 x 10 = 1.0 mmol
    $[I_3^-] = 1.0\ mmol/100mL = 0.0100\ M$
    $[I^-] = [(5.0 \times 90) - 3 \times 1.0]\ mmol/100\ mL = 4.5\ M$
    $E = 0.536 - (0.059/2) \log ([I^-]^3/[I_3^-])$
    $= 0.536 - (0.059/2) \log [(4.5)^3/(0.010)] = 0.419\ V$

18. $E = 0.68 - (0.059/2) \log ([PtCl_4^{2-}][Cl^-]^2/[PtCl_6^{2-}])$
    We can assume $[Cl^-] = 3.0\ M$ since the amount used in complexing the Pt is small.
    $\therefore\ E = 0.68 - (0.059/2) \log [(0.025)(3.0)^2/(0.015)] = 0.65\ V$

19. $UO_2^{2+} + 4H^+ + 2e^- = U^{4+} + 2H_2O$ $\quad\quad\quad\quad E^0 = 0.334$ V

$\underline{V^{3+} + e^- = V^{2+} \quad\quad\quad\quad\quad\quad\quad\quad\quad E^0 = -0.255 \text{ V}}$

$UO_2^{2+} + 2V^{2+} + 4H^+ = U^{4+} + 2V^{3+} + 2H_2O \quad E^0_{cell} = 0.589$ V

Assuming 1 mL each, we have 0.1 mmol of UO_2^{2+}, 0.0 mmol of V^{2+}, and 0.2 mmol H_2SO_4 (0.4 mmol H^+) in 2 mL. Each millimole of V^{2+} will react with only ½ mmol of UO_2^{2+} and 2 mmol H^+, so we have an excess of UO_2^{2+}.

mmol UO_2^{2+} = 0.100 − 0.050 = 0.050 mmol left/2 mL

mmol U^{4+} produced = 0.050 mmol/2 mL

mmol H^+ = 0.40 − 0.20 = 0.20 mmol left/2 mL = 0.10 M

Volumes cancel for the uranium species, so we can use millimoles.

∴ E = 0.334 - (0.059/2) log ([U^{4+}]/[UO_2^{2+}][H^+]^4)

= 0.334 - (0.059/2) log [(0.050)/(0.050)(0.10)^4] = 0.216 V

20. Since the $PtCl_6^{2-}/PtCl_4^{2-}$ potential is the more positive, $PtCl_6^{2-}$ will oxidize V^{2+}, or subtracting the second half-reaction (multipied by 2) from the first to give a positive

E_{cell}, the reaction is:

$PtCl_6^{2-} + 2V^{2+} = PtCl_4^{2-} + 2V^{3+} + 2Cl^-$ $\quad\quad E_{cell} = 0.68 - (-0.255)$ V = 0.94V

21. (a) $E_{cell} = E_{cathode} - E_{anode} = 1.20 - (0.059/10) \log ([I_2]/[IO_3^-]^2[H^+]^{12} - 0.5455 +$

(0.059/2) log ([I^-]^3/I_3^-])

$E_{cell} = 1.20 - (0.059/10) \log [(0.0100)/(0.100)(0.100)^{12}]$ −

0.54 + (0.059/2) log [(0.100)^3/(0.0100)] = 0.57 V

(b) $E_{cell} = 0.334 - (0.059/2) \log ([U^{4+}]/[UO_2^{2+}][H^+]^4) - 0.222 + 0.059 \log [Cl^-]$

$E_{cell} = 0.334 - (0.059/2) \log [(0.0500)/(0.200)(1.00)^4] - 0.222 + 0.059 \log (0.100)$

= 0.071V

(c) $E_{cell} = 1.51 - (0.059/5) \log ([Mn^{2+}]/[MnO_4^-][H^+]^8) -$

1.25 + (0.059/2) log ([Tl^+]/[Tl^{3+}])

$E_{cell} = 1.51 - (0.059/5) \log [(0.100)/(0.0100)(1.0 \times 10^{-2})^8]$ -

1.25 + (0.059/2) log ((0.0100)/(0.0100) = 0.09 V

22. $VO_2^+ + 2H^+ + e^- = VO^{2+} + H_2O$

$E^0 = 1.00$ V $\quad\quad\quad\quad\quad\quad\quad\quad$ (1)

$UO_2^{2+} + 4H^+ + 2e^- = U^{4+} + 2H_2O$

$E^0 = 0.334$ V $\quad\quad\quad\quad\quad\quad\quad$ (2)

Subtracting (2) from (1) (multiplied by 2) gives a positive cell potential and, hence, the spontaneous reaction:

$2VO_2^+ + U^{4+} = 2VO^{2+} + UO_2^{2+}$ $\quad\quad\quad\quad\quad\quad E^0_{cell} = 1.00 - 0.334 = 0.67$ V

23. PFP

$Cu^{2+} + 2e^- \Leftrightarrow Cu(s)$ $E_+^o = 0.337$ V

$Ni^{2+} + 2e^- \Leftrightarrow Ni(s)$ $E_-^o = -0.250$ V

$E_+ = 0.337$ V $- (0.05916$ V$/2)$ $\log(1/0.0030) = 0.262$ V

$E_- = -0.250$ V $- (0.05916$ V$/2)$ $\log(1/0.0020) = -0.330$ V

$E = E_+ - E_- = 0.262$ V $- (-0.330$ V$) = 0.592$ V

24. PFP

(a) Cu: 0.286 V; Ag: 0.740 V; cathode.

$$E_{Cu^{2+}/Cu} = E^0_{Cu^{2+}/Cu} - \frac{0.05916}{2}\log\frac{1}{[Cu^{2+}]} = 0.339 - \frac{0.05916}{2}\log\frac{1}{0.0167} = 0.286(V)$$

$$E_{Ag^+/Ag} = E^0_{Ag^+/Ag} - \frac{0.05916}{1}\log\frac{1}{[Ag^+]} = 0.799 - \frac{0.05916}{1}\log\frac{1}{0.100} = 0.740(V)$$

(b) 0.706 V; cathode.

$$mol_{KCl} = \frac{0.271g}{74.55g/mol} = 0.003635mol$$

$$mol_{Ag^+} = 50.0mL \times 0.100M = 0.00500mol$$

$$Ag^+ + Cl^- = AgCl_{(s)}$$

After reaction, $$[Ag^+] = \frac{0.00500mol - 0.003635mol}{50.0mL} = 0.0273M$$

$$E_{Ag^+/Ag} = E^0_{Ag^+/Ag} - \frac{0.05916}{1}\log\frac{1}{[Ag^+]} = 0.799 - \frac{0.05916}{1}\log\frac{1}{0.0273} = 0.706(V)$$

(c) 0.277 V; anode.

$$mol_{KCl} = \frac{0.812g}{74.55g/mol} = 0.01089mol$$

$$mol_{Ag^+} = 50.0mL \times 0.100M = 0.00500mol$$

$$Ag^+ + Cl^- = AgCl_{(s)}$$

After reaction, $$[Cl^-] = \frac{0.01089mol - 0.00500mol}{50.0mL} = 0.1178M$$

$$[Ag^+] = \frac{K_{sp}}{[Cl^-]} = \frac{1.8\times10^{-10}}{0.1178} = 1.528\times10^{-9}(M)$$

$$E_{Ag^+/Ag} = E^0_{Ag^+/Ag} - \frac{0.05916}{1}\log\frac{1}{[Ag^+]} = 0.799 - \frac{0.05916}{1}\log\frac{1}{1.528\times10^{-9}} = 0.277(V)$$

CHAPTER 13 POTENTIOMETRIC ELECTRODES AND POTENTIOMETRY

1. The liquid junction potential is the potential existing at a boundary between two dissimilar solutions (e.g., at a solution–salt bridge interface). It is due to unequal diffusion of ions across the boundary, which results in a net charge built up on each side of the boundary (a potential). It can be minimized by adding a high concentration of an electrolyte on one side of the boundary, whose cation and anion diffuse at nearly equal rates (e.g, saturated KCl).

2. The generally accepted theory of the glass electrode response to protons is the result of migration of solution protons to the hydrated glass surface gel containing low activity protons from the –SiOH groups, building up a microscopic layer of positive charges, the boundary potential. Other theories are offered based on chemisorption and charge separation (Pungor) or capacitance (Cheng).

3. The alkaline error is the result of competition between H^+ and Na^+ ions for the potential determining mechanism. At high pH, response to Na^+ becomes appreciable, causing the pH reading to be "too acidic". The acid error occurs in very acidic solution and is a result of the decrease of water activity, causing a positive error in the pH reading.

4. The general construction of ion selective electrodes is: internal reference electrode/internal filling solution/membrane. The main difference in the electrodes lies in differences in their membranes. The four major types of membranes include glass membranes (useful for monovalent cations), precipitate impregnated membranes (useful for anions and some cations), solid state (crystalline) membranes (useful for halides, especially fluoride, sulfide, cyanide, and some cations), and liquid-liquid membranes (useful for several cations and anions, especially Ca^{2+}, nitrate, perchlorate, and chloride).

5. It is a measure of the selectivity of the membrane response of an ion selective electrode for one ion over another and can be used in the Nernst equation to predict the response due to an interfering ion. It can be estimated by measuring the potential of two different solutions containing different concentratons of the two ions to which the electrode responds (separate solution method), or by measuring the potential of solution of mixtures of the ions (mixed solution method).

6. A crown ether is a neutral cyclic ether containing several oxygens in a ring or cage that is of appropriate size to incorporate and complex metal ions of certain sizes. An 18-crown-6 ether contains 6 oxygens in an 18-membered ring.

7. The Nicolsky equation describes the potential of an electrode that responds to two (or more) ions, and is given by Equation 13.46..

8. $E^0_{AgBr,Ag} = E^0_{Ag^+,Ag} + 0.05916 \log K_{sp}$
   $0.073 = 0.799 + 0.05916 \log K_{sp}$

$\log K_{sp} = -12.2_7$
$K_{sp} = 10^{-12.2}{}_7 = 10^{-13} \times 10^{0.7}{}_3 = 5.4 \times 10^{-13}$

9. $E_{cell} = E^0{}_{Ag^+,Ag} - 0.05916 \log (1/a_{Ag^+}) - E_{S.C.E.}$
    At the end point, a saturated solution of AgSCN exists.
    $\therefore [Ag^+] = \sqrt{K_{sp}} = 1.00 \times 10^{-6} M$
    $0.202 = E^0{}_{Ag^+,Ag} - 0.05916 \log [1/(1.00 \times 10^{-6})] - 0.242$
    $E^0{}_{Ag^+,Ag} = 0.799 V$

10. (a) (1) $Ag = Ag^+ + e^-$ $E^0 = 0.799$
           $Fe^{3+} + e^- = Fe^{2+}$ $E^0 = -0.771$
       (2) $Ag/Ag^+, Fe^{3+}, Fe^{2+}/Pt$
       (3) $E^0{}_{cell} = E^0{}_{right} - E^0{}_{left} = 0.771 - 0.799 = -0.028 V$
    Since the potential is negative, the reaction will not go as written under standard conditions.
       (4) Ag electrode = -   Pt electrode = +

    (b) (1) $VO_2^+ + 2H^+ + e^- = VO^{2+} + H_2O$ $E^0 = 1.000$
           $V^{3+} + H_2O = VO^{2+} + 2H^+ + e^-$ $E^0 = 0.361$
       (2) $Pt/V^{3+}, VO_2^+, VO^{2+}/Pt$
       (3) $E^0{}_{cell} = 1.000 - 0.361 = 0.639 V$
       (4) left Pt electrode = -   right Pt electrode = +

    (c) (1) $Ce^{4+} + e^- = Ce^{3+}$ $E^0 = 1.61$
           $Fe^{2+} = Fe^{3+} + e^-$ $E^0 = 0.771$
       (2) $Pt/ Fe^{2+}, Fe^{3+}, Ce^{4+}, Ce^{3+}/Pt$
       (3) $E^0{}_{cell} = 1.61 - 0.77 = 0.84 V$
       (4) Left Pt electrode = -   right Pt electrode = +

11. (a) (1) $H_2 = 2H^+ + 2e^-$ $E^0 = 0.000 V$
           $\underline{Cl_2 + 2e^- = 2Cl^-}$ $E^0 = 1.359 V$
           $Cl_2 + H_2 = 2HCl$

       (2) $E_{cell} = E_{right} - E_{left}$
           $= 1.359 - (0.05916/2) \log ([Cl^-]^2/p_{Cl_2}) + (0.05916/2) \log (p_{H_2}/[H^+]^2)$
           $= 1.359 - (0.05916/2 \log [(0.5)^2/(0.20)] + (0.05916/2) \log [(0.2)/(0.5)^2]$
           $= 1.353 V$

    (b) (1) $Fe^{2+} = VO^{2+} + e^-$ $E^0 = 0.771 V$
           $\underline{VO_2^+ + 2H^+ + e^- = VO^{2+} + H_2O}$ $E^0 = 1.000 V$
           $VO_2^+ + Fe^{2+} + 2H^+ = VO^{2+} + Fe^{3+} + H_2O$

       (2) $E_{cell} = 1.000 - 0.05916 \log [VO^{2+}]/([VO_2^+][H^+]^2)$
           $= 0.771 + 0.05916 \log ([Fe^{2+}]/[Fe^{3+}])$
           $= 1.000 - 0.05916 \log (0.002)/[(0.001)(0.1)^2]$
           $- 0.771 + 0.05916 \log [(0.005)/(0.05)] = 0.034 V$

12. PFP

Potential of SCE is 0.242 vs. NHE. Therefore, the standard reduction potential of the PVF film is,

$$E^0_{PVF^+/PVF} = 0.296 + 0.242 = 0.538 \, (\text{V } vs. \text{ NHE})$$

For the AuCl$_4^-$/Au pair,

$$AuCl_4^- + 3e^- = Au + 4Cl^-$$

$$E^0_{AuCl_4^-/Au} = 1.002 \, (\text{V } vs. \text{ NHE})$$

The reaction mentioned will be

$$3PVF + AuCl_4^- = 3PVF^+ + Au + 4Cl^-$$

$$\Delta E^0 = E^0_{cathode} - E^0_{anode} = 1.002 - 0.538 = 0.464 \, (\text{V})$$

$$\Delta G^0 = -nFE^0 = -RT \ln K$$

$$K = e^{\frac{nFE^0}{RT}} = \exp\left(\frac{3 \times 96485.3 \times 1.002}{8.31 \times 298.15}\right) = 6.9 \times 10^{70}$$

It is very likely that PVF is oxidized and AuCl$_4^-$ is reduced to produce Au particles. This is because the above reaction has a very large value of equilibrium constant. Though the concentration of KAuCl$_4$ is as small as 2.0 mM, the reduction of KAuCl$_4$ will still happen.

13. (a) $E_{vs \, SCE} = 1.087 - 0.242 = 0.845$ V

(b) $E_{vs \, SCE} = 0.22 - 0.242 = -0.020$ V

(c) $E_{vs \, SCE} = -0.255 - 0.242 = -0.497$ V

14. $E_{Pt} = 0.771 - 0.05916 \log [(0.0500)/(0.00200)] = 0.688$ V
$E_{cell} = E_{Pt} - E_{SCE} = 0.688 - 0.242 = 0.446$ V

15. (a) When the I$^-$ is all titrated, the volume is doubled and [Cl$^-$] = 0.050 M when AgCl begins to precipitate:
$[Ag^+] = K_{sp(AgCl)}/[Cl^-] = (1.0 \times 10^{-10})/(0.050) = 2.0 \times 10^{-9}$ M
$[I^-] = K_{sp(AgCl)}/[Ag^+] = (1 \times 10^{-16})/(2.0 \times 10^{-9}) = 5 \times 10^{-8}$ M
% I$^-$ remaining = $[(5 \times 10^{-8} \, M \times 100 \, \text{mL})/(1.0 \times 10^{-1} \, M \times 50 \, \text{mL})] \times 100\%$
= 1×10^{-4} %

(b) $E = E^0_{Ag^+,Ag} - .05916 \log (1/a_{Ag^+}) - E_{S.C.E.} - 0.799 - 0.05916 \log [1/(2.0 \times 10^{-9})]$
$- 0.242 = 0.042$ V

At the theoretical end point, we have a saturated solution of AgI, and:

$[Ag^+] = \sqrt{K_{sp(AgI)}} = \sqrt{1 \times 10^{-16}} = 1 \times 10^{-8}$ M
$E = 0.799 - 0.05916 \log [1/(1 \times 10^{-8})] - 0.242 = 0.084_4$ V

(d) At the end point, we have a saturated solution of AgCl, and:

$$[Ag^+] = \sqrt{K_{sp(AgCl)}} = \sqrt{1.0x10^{-10}} = 1.0 \times 10^{-5}\ M$$

$$E = 0.799 - 0.05916 \log [1/(1.0 \times 10^{-5})] - 0.242 = 0.261\ V$$

16. $E = E^0_{Hg^{2+},Hg} - (2.303\ RT/2F) \log (1/a_{Hg^{2+}})$

$K_{f(Hg\text{-}EDTA)} = (a_{(Hg\text{-}EDTA)}/a_{Hg^{2+}}.a_{EDTA^{2-}})$

$a_{Hg^{2+}} = [1/K_{f(Hg\text{-}EDTA)}].(a_{Hg\text{-}EDTA}/a_{EDTA^{2-}})$

Substituting this in the equation,

$E = E^0_{Hg^{2+},Hg} - (2.303\ RT/2F) \log K_{f(Hg\text{-}EDTA)} - (2.303\ RT/2F\ \log (a_{EDTA^{2-}}/a_{Hg\text{-}EDTA})$

$K_{f(M\text{-}EDTA)} = (a_{M\text{-}EDTA})/(a_M{}^{n+}.a_{EDTA^{2-}})$

$a_{EDTA^{2-}} = [1/K_{f(M\text{-}EDTA)}](a_{M\text{-}EDTA}/a_M{}^{n+})$

Substituting this in the equation,

$E = E^0_{Hg^{2+},Hg} - (2.302\ RT/2F) \log [K_{fHg\text{-}EDTA}/K_{f(M\text{-}EDTA)}) - 2.303\ RT/F \log (a_{M\text{-}EDTA}/a_{Hg\text{-}EDTA})$
$- (2.303\ RT/2F) \log (1/a_M{}^{n+})$

17. From the appendix, the potential of the calomel electrode in 1 M KCl is +0.282 V.

$E_{vs\ S.C.E.} = E_{ind} - E_{S.C.E.}$

$-0.465\ V = E_{ind} - 0.242\ V$

$E_{ind} = -0.223\ V$

$E_{vs\ N.C.E.} = E_{ind} - E_{N.C.E} = -0.223\ V - 0.282\ V = -0.505\ V$

That is, since the N.C.E. is 0.040 V more positive, the indicator electrode will be 0.040 V more negative relative to this electrode.

18. (a) About 0.02 pH unit. It is limited by the accuracy to which the pH of the standardizing solution is known, which is limited in turn by the certainy to which the activity of a single ion can be calculated (that ion is Cl⁻ in the reference AgCl/Ag electrode used in standardizing measurements), and by the residual liquid junction potential.

$E = 0.059\ \Delta pH = 0.059\ (0.02) = 0.0012\ V = 1.2\ mV$

For n = 1, the error is 4% per mV. Therefore, the error in hydrogen ion activity is 4.8%.

(b) About 0.002 pH unit.

$\Delta E = (0.059)(0.002) = 0.0012\ V = 0.12\ mV$

$(4\%/mV)(0.12\ mV) = 0.48\%$ variation in hydrogen ion activity

19. $E = k - 0.05916\ pH$

$0.395 = k - (0.05916)(7.00)$

$k = 0.809$

(a) $pH = (k - E)/(0.05916) = (0.809 - 0.467)/(0.05916) = 5.78$

(b) $pH = (0.089 - 0.209)/(0.05916) = 10.14$

(c) $pH = (0.089 - 0.080)/(0.05916) = 12.32$

(d) $pH = (0.089 + 0.013)/(0.05916) = 13.89$

20. (a) pH = 3.00

$E_{cell} = E_{H^+,H2} - E_{S.C.E.} = -0.05916 \log (1/a_{H^+}) - 0.242$

$= -0.05916pH - 0.242 = -0.05916(3.00) - 0.242 = -0.419$ V

(i.e., the potential of the indicating electrode is $-$ 0.419 V relative to the reference electrode; or E = 0.49 V with the reference electrode the more positive.)

(b) $H^+ = \sqrt{K_a[HOAc]} = \sqrt{(1.75x10^{-5})(1.00x10^{-3})} = 1.32 \times 10^{-4}$

pH = 3.88

$E_{cell} = (0.05916)(3.88) - 0.242 = -0.472$ V

(ΔE = 0.472 V with the reference electrode the more positive)

(c) pH = pKa + log ([OAc⁻]/HOAc) = 4.76 + log 1.00 = 4.76

$E_{cell} = (0.05916)(4.76) - 0.242 = -0.524$ V

(ΔE = 0.524 V with the reference electrode the more positive)

21. $E_{cell} = E_{ind} - E_{rev} = E_{ind} - 0.242$

In this case, the indicating electrode potential is less positive than that of the S.C.E.

$-0.205 = 0.699 - (0.05916/2) \log [a_{HQ}/(a_Q a_{H^+}{}^2)] - 0.242$

Since the ratio of a_{HQ}/a_Q is unity,

$-0.662 = -(0.05916/2) \log (1/a_{H^+}{}^2) = 0.05916 \log a_{H^+} = -0.05916$ pH

pH = 11.2

22. In a mixture of $A^+ + B^+$, the electrode potential is given by:

$E_{AB} = k_A + S \log (a_A{}^+ + K_{AB}a_B{}^+)$ (1)

In a solution containing only B^+, it is given by:

$E_B = k_B + S \log a_B{}^+$ (2)

Also, in a solution containing only B^+, (1) reduces to:

$E_B = k_A + S \log K_{AB}a_B{}^+ = k_A + S \log K_{AB} + S \log a_B{}^+$ (3)

Hence, (2) and (3) are equal:

$k_B + S \log a_B{}^+ = k_A + S \log K_{AB} + S \log a_B{}^+$; $\log K_{AB} = (k_B - k_A)/S$ (4)

23. The ionic strength of the KCl standard is 0.0050. From Equation 6.20 in Chapter 6,

$-\log f_i = 0.51 (1)^2 \sqrt{0.0050} /(1 + \sqrt{0.0050})$

$f_i = 0.93$

$a_K{}^+ = 0.0050 \, M \times 0.93 = 0.0046 \, M$

Standard:

-18.3 mV $= k + 59.2 \log (0.0046)$

$k = +120.1$ mV

Sample:

$E = k + 59.2 \log (a_{K^+} + K_{KCs}a_{Cs^+})$

$20.9 = +120.2 + 59.2 \log (a_{K^+} + 1 \times 0.0060)$

$-1.68 = \log (a_{K^+} + 0.0060)$

$a_{K^+} = 0.015 \, M$

24. Calculate S:

$-108.6 = k - S \log (0.0500)$ (1)

$\underline{-125.2 = k - S \log (0.0100)}$ (2)

(1) – (2): $16.6 = -S \log (0.0050) + S \log (0.0100)/(0l0050)$

$S = 55.1$

Calculate k. Use either (1) or (2):

$-125.2 \, mV = k - 55.1 \log (0.0100)$

$k = -235.4 \, mV$

For the sample:

$-119.6 = -235.4 - 55.1 \log (NO_3^-)$

$[NO_3^-] = 0.0079 \, M$

25. For ClO_4^- :

$-27.2 = k_{ClO4^-} - 59.2 \log (0.00100); \ k_{ClO4^-} = -204.8 \, mV$

For I^-:

$+32.8 = k_{I^-} - 59.2 \log (0.0100) = k_{ClO4^-} - 59.2 \log [K_{ClO4^-,I^-} (0.0100)]$

$k_{I^-} = -85.6 \, mV$

$\log K_{ClO4^-,I^-} = (k_{ClO4^-} - k_{I^-})/59.2 = (-204.8 + 85.6)/59.2 = -2.01$

$K_{ClO4^-,I^-} = 0.0097$

For sample:

$-15.5 = -204.8 - 59.2 \log ([ClO_4^-] + 9.7 \times 10^{-3} (0.015))$

$-3.20 = \log ([ClO_4^-] + 1.5 \times 10^{-4})$

$[ClO_4^-] = 6.3 \times 10^{-4} - 1.5 \times 10^{-4} = 4.8 \times 10^{-4} \, M$

26. (a) $k_{Na} = E_{Na} - 59.2 \log a_{Na+} = 113.0 - 59.2 \log 1.00 \times 10^{-1} = 172.2 \, mV$

    $k_K = 67.0 - 59.2 \log 1.00 \times 10^{-1} = 126.2 \, mV$

    $\log K_{NaK} = (k_K - k_{Na})/59.2 = (126.2 - 172.2)/59.2 = -0.777$

    $K_{NaK} = 0.16$

(b) $E_{NaK} = k_{Na} + 59.2 \log (a_{Na^+} + K_{NaK}a_{K^+})$

    $= 172.2 + 59.2 \log (1.00 \times 10^{-3} + 0.167 \times 1.00 \times 10^{-2}) = 18.9 \, mV$

27. $+237.8 = k + 56.1 \log [2.00 \times 10^{-4} + K_{AB} (1.00 \times 10^{-3})]$ (1)

$\underline{+253.6 = k + 56.1 \log [4.00 \times 10^{-4} + K_{AB} (1.00 \times 10^{-3})]}$ (2)

(2) – (1):

$20.4 = 56.1 \log [k + 56.1 \log [4.00 \times 10^{-4} + K_{AB} (1.00 \times 10^{-3})]/[2.00 \times 10^{-4} + K_{AB} (1.00 \times 10^{-3})]$

1.91 $[2.00 \times 10^{-4} + K_{AB} (1.00 \times 10^{-3})] = 4.00 \times 10^{-4} + K_{AB} (1.00 \times 10^{-3})$
$K_{AB} = 0.020$

28. $-\log_{NaK} = E_{Na} - E_K/S = E/S = 175.5/58.1 = 3.02$
   $K_{NaK} = 10^{-3.02} = 10^{-4} \times 10^{.98} = 9.5 \times 10^{-4}$

29. From Equation 13.52, the electrode responds equally to the two ions when the deviation from the experimental curve is $(0.301)(57.8)/1 = 17.4$ mV.

   From Equation 13.50,
   $K_{KNa} = a_K/a_{Na} = 0.015$ mM/140 mM $= 1.0_7 \times 10^4$

CHAPTER 14   REDOX AND POTENTIOMETRIC TITRATIONS

1. Starch indicator for iodometric and iodimetric titrations.
    Self indicator for highly colored titrants, e.g., $KMnO_4$.

    Redox indicators for other titrations. The indicator is itself a weak oxidizing or reducing agent whose oxidized and reduced forms are different colors. E^0_{ind} should be near the equivalence point potential.

2. In iodimetry, a reducing agent is titrated with a standard solution of I_2 to a blue starch endpoint. In iodometry, an oxidizing agent is reacted with an excess of I^- to form an equivalent amount of I_2, which is then titrated with a standard solution of $Na_2S_2O_3$ to a starch endpoint (disappearance of bule I_2-starch color).

3. Many oxidizing agents that react with iodide consume protons in the reaction, and the equilibrium lies far to the right only in acid solution. Conversely, most reducing agents that react with I_2 liberate protons, and the solution must be near neutrality for the equilibrium to lie far to the right. Also, in iodimetry, the pH is kept near neutrality to minimize acid hydrolysis of the starch and air oxidation of the iodide produced.

4. No. It is slightly beyond the equivalence point, because an excess of permanganate must be added to impart a pink color to the solution. A blank titration can be used to correct for the excess.

5. It contains $MnSO_4$ and H_3PO_4. The Mn^{2+} is added to decrease the potential of the MnO_4^- /Mn^{2+} couple, so that permanganate will not oxidize Cl^-. The H_3PO_4 is added to complex the Fe^{3+} and decrease the potential of the Fe^{3+}/Fe^{2+} couple and hence sharpen the endpoint. The iron(III) complex is also colorless, which makes the endpoint easier to see.

6. We use the procedure in which individual half-reactions are balanced without knowledge of the oxidation states. The O's are balanced with H_2O, the H's with H^+ (followed by neutralization with OH^- if alkaline), and finally, the charges with electrons.

    (a) (1) $IO_3^- = \frac{1}{2} I_2$
        (2) $IO_3^- = \frac{1}{2} I_2 + 3H_2O$
        (3) $IO_3^- + 6H^+ = \frac{1}{2} I_2 + 3H_2O$
        (4) $IO_3^- + 6H^+ + 5e^- = \frac{1}{2} I_2 + 3H_2O$

        (1) $I^- = \frac{1}{2} I_2$
        (2) $I^- = \frac{1}{2} I_2 + e^-$

        $IO_3^- + 6H^+ + 5e^- = \frac{1}{2} I_2 + 3H_2O$
        $5(I^- = \frac{1}{2} I_2)$

        _____

        $IO_3^- + 5I^- + 6H^+ = 3I_2 + 3H_2O$

(b) (1) $Se_2Cl_2 = 2H_2SeO_3 + 2Cl^-$
    (2) $Se_2Cl_2 + 6H_2O = 2H_2SeO_3 + 2Cl^-$
    (3) $Se_2Cl_2 + 6H_2O = 2H_2SeO_3 + 2Cl^- + 8H^+$
    (4) $Se_2Cl_2 + 6H_2O = 2H_2SeO_3 + 2Cl^- + 8H^+ + 6e^-$

    (1) $Se_2Cl_2 = 2Se + 2Cl^-$
    (2) $Se_2Cl_2 + 2e^- = 2Se + 2Cl^-$

    $Se_2Cl_2 + 6H_2O = 2H_2SeO_3 + 2Cl^- + 8H^+ + 6e^-$
    $3(Se_2Cl_2 + 2e^- = 2Se + 2Cl^-)$
    _____

    $4Se_2Cl_2 + 6H_2O = 2H_2SeO_3 + 6Se + 8Cl^- + 8H^+$

(c) (1) $H_3PO_3 = H_3PO_4$
    (2) $H_3PO_3 + H_2O = H_3PO_4$
    (3) $H_3PO_3 + H_2O = H_3PO_4 + 2H^+$
    (4) $H_3PO_3 + H_2O = H_3PO_4 + 2H^+ + 2e^-$

    (1) $2HgCl_2 = Hg_2Cl_2 + 2Cl^-$
    (2) $2HgCl_2 + 2e^- = Hg_2Cl_2 + 2Cl^-$

    $H_3PO_3 + H_2O = H_3PO_4 + 2H^+ + 2e^-$
    $2HgCl_2 + 2e^- = Hg_2Cl_2 + 2Cl^-$
    _____

    $H_3PO_3 + 2HgCl_2 + H_2O = H_3PO_4 + Hg_2Cl_2 + 2H^+ + 2Cl^-$

7. (a) (1) $MnO_4^- = MnO_2$
    (2) $MnO_4^{2-} = MnO_2 + 2H_2O$
    (3) $MnO_4^{2-} + 4H^+ = MnO_2 + 2H_2O$
        $MnO_4^{2-} + 2H_2O = MnO_2 + 4OH^-$
    (4) $MnO_4^{2-} + 2H_2O + 2e^- = MnO_2 + 4OH^-$

    (1) $MnO_4^{2-} = MnO_4^-$
    (2) $MnO_4^{2-} = MnO_4^- + e^-$

    $MnO_4^{2-} + 2H_2O + 2e^- = MnO_2 + 4OH^-$
    $2(MnO_4^{2-} = MnO_4^- + e^-)$
    _____

    $3MnO_4^{2-} + 2H_2O = MnO_2 + 2MnO_4^- + 4OH^-$

(b) (1) $MnO_4^- = Mn^{2+}$
    (2) $MnO_4^- = Mn^{2+} + 4H_2O$
    (3) $MnO_4^- + 8H^+ = Mn^{2+} + 4H_2O$
    (4) $MnO_4^- + 8H^+ + 5e^- = Mn^{2+} + 4H_2O$

    (1) $H_2S = S$
    (2) $H_2S = S + 2H^+$

(3) $H_2S = S + 2H^+ + 2e^-$

$2(MnO_4^- + 8H^+ + 5e^- = Mn^{2+} + 4H_2O)$

$5(H_2S = S + 2H^+ + 2e^-)$

$$\overline{2MnO_4^- + 5H_2S + 6H^+ = 2Mn^{2+} + 5S + 8H_2O}$$

(c) (1) $2SbH_3 = H_4Sb_2O_7$

(2) $2SbH_3 + 7H_2O = H_4Sb_2O_7$

(3) $2SbH_3 + 7H_2O = H_4Sb_2O_7 + 16H^+$

(4) $2SbH_3 + 7H_2O = H_4Sb_2O_7 + 16H^+ + 16e^-$

(1) $Cl_2O = 2Cl^-$

(2) $Cl_2O = 2Cl^- + H_2O$

(3) $Cl_2O + 2 H^+ = 2Cl^- + H_2O$

(4) $Cl_2O + 2 H^+ + 4e^- = 2Cl^- + H_2O$

$2SbH_3 + 7H_2O = H_4Sb_2O_7 + 16H^+$

$4(Cl_2O + 2 H^+ + 4e^- = 2Cl^- + H_2O)$

$$\overline{2SbH_3 + 4Cl_2O + 3H_2O = H_4Sb_2O_7 + 8Cl^- + 8H^+}$$

(d) (1) $FeS = Fe^{3+} + S$

(2) $FeS = Fe^{3+} + S + 3e^-$

(1) $NO_3^- = NO_2$

(2) $NO_3^- = NO_2 + H_2O$

(3) $NO_3^- + 2H^+ = NO_2 + H_2O$

(4) $NO_3^- + 2H^+ + e^- = NO_2 + H_2O$

$FeS = Fe^{3+} + S + 3e^-$

$3(NO_3^- + 2H^+ + e^- = NO_2 + H_2O)$

$$\overline{FeS + 3NO_3^- + 6H^+ = Fe^{3+} + S + 3NO_2 + 3H_2O}$$

(e) (1) $Al = AlO_2^-$

(2) $Al + 2H_2O = AlO_2^-$

(3) $Al + 2H_2O = AlO_2^- + 4H^+$

$Al + 4OH^- = AlO_2^- + 2H_2O$

(4) $Al + 4OH^- = AlO_2^- + 2H_2O + 3e^-$

(1) $NO_3^- = NH_3$

(2) $NO_3^- = NH_3 + 3H_2O$

(3) $NO_3^- + 9H^+ = NH_3 + 3H_2O$

$NO_3^- + 6H_2O = NH_3 + 9OH^-$

(4) $NO_3^- + 6H_2O + 8e^- = NH_3 + 9OH^-$

$8(Al + 4OH^- = AlO_2^- + 2H_2O + 3e^-)$
$3(NO_3^- + 6H_2O + 8e^- = NH_3 + 9OH^-)$

———————————————————————————————

$8Al + 3NO_3^- + 5OH^- + 2H_2O = 8AlO_2^- + 3NH_3$

(f) (1) $FeAsS = Fe^{3+} + AsO_4^{3-} + SO_4^{2-}$
    (2) $FeAsS + 8H_2O = Fe^{3+} + AsO_4^{3-} + SO_4^{2-}$
    (3) $FeAsS + 8H_2O = Fe^{3+} + AsO_4^{3-} + SO_4^{2-} + 16H^+$
    (4) $FeAsS + 8H_2O = Fe^{3+} + AsO_4^{3-} + SO_4^{2-} + 16H^+ + 14e^-$

    (1) $ClO_2 = Cl^-$
    (2) $ClO_2 = Cl^- + 2H_2O$
    (3) $ClO_2 + 4H^+ = Cl^- + 2H_2O$
    (4) $ClO_2 + 4H^+ + 5e^- = Cl^- + 2H_2O$

    $5(FeAsS + 8H_2O = Fe^{3+} + AsO_4^{3-} + SO_4^{2-} + 16H^+ + 14e^-)$
    $14(ClO_2 + 4H^+ + 5e^- = Cl^- + 2H_2O)$

    ———————————————————————————————

    $5FeAsS + 14ClO_2 + 12H_2O = 5Fe^{3+} + 5AsO_4^{3-} + 5SO_4^{2-} + 14Cl^- + 24H^+$

(g) (1) $K_2NaCo(NO_2)_6 = 2K^+ + Na^+ + Co^{3+} + 6NO_3^-$
    (2) $K_2NaCo(NO_2)_6 = 2K^+ + Na^+ + Co^{3+} + 6NO_3^-$
    (3) $K_2NaCo(NO_2)_6 = 2K^+ + Na^+ + Co^{3+} + 6NO_3^- + 12H^+$
    (4) $K_2NaCo(NO_2)_6 = 2K^+ + Na^+ + Co^{3+} + 6NO_3^- + 12H^+ + 12e^-$

    $5(K_2NaCo(NO_2)_6 = 2K^+ + Na^+ + Co^{3+} + 6NO_3^- + 12H^+ + 12e^-)$
    $12(MnO_4^- + 8H^+ + 5e^- = Mn^{2+} + 4H_2O)$

    ———————————————————————————————

    $5K_2NaCo(NO_2)_6 + 12MnO_4^- + 36H^+ = 10K^+ + 5Na^+ + 5Co^{3+} + 30NO_3^- + 12Mn^{2+} + 18H_2O$

8. $Tl^+ + 2Co^{3+} = Ti^{3+} + 2Co^{2+}$
    Equivalent amounts of Tl^+ and Co^{3+} are added. Thre are no polynuclear species nor H^+.
    $\therefore E = (2 \times 0.77 + 1 \times 1.1842)/(2 + 1) = 1.12_7$ V

9. $6Fe^{2+} + Cr_2O_7^{2-} + 14H^+ = 6Fe^{3+} + 2Cr^{3+} + 7H_2O$

    Before the enpoint, the half cell
    $Fe^{3+} + e^- = Fe^{2+}$ is used for the potential calculation, since we know the concentration of each species.

    After the endpoint, the half cell
    $Cr_2O_7^{2-} + 14H^+ + 6e^- = 2Cr^{3+} + 7H_2O$ is used.

    We start with 5.00 mmol Fe^{2+} and 50.0 mmol H^+
    10 mL: mmol Fe^{3+} = 1.00
              mmol Fe^{2+} = 4.00

mmol Cr^{3+} = 0.334
mmol H^+ = 47.7
E = 0.771 −0.059 log (4.00/1.00) = 0.735 V

25.0 mL: mmol Fe^{3+} = 2.50
mmol Fe^{2+} = 2.50
mmol Cr^{3+} = 0.835
mmol H^+ = 44.2
E = 0.771 −0.059 log (2.50/2.50) = 0.771 V

50.0 mL: mmol Fe^{3+} = 5.00 − x ≈ 5.00
mmol Fe^{2+} = x
mmol Cr^{3+} = 1.67 − 1/3x ≈ 1.67; M = 0.0167 mmol/mL
mmol $Cr_2O_7^{2-}$ = 1/6x; M = (1/6 x/100) mmol/mL
mmol H^+ = 38.3 + 14/6 x ≈ 38.4; M = 0.383 mmol/mL
Calculate K_{eq}:
0.771 − (0.059/6) log ($[Fe^{2+}]^6/[Fe^{3+}]^6$)
= 1.33 − (0.059/6) log ($[Cr^{3+}]^2/[Cr_2O_7^{2-}][H^+]^{14}$)
-log K_{eq} = - log ($[Fe^{3+}]^6[Cr^{3+}]^2$)/($[Fe^{2+}]^6[Cr_2O_7^{2-}][H^+]^{14}$)
= (0.771 − 1.33)/(0.059/6)
K_{eq} = 7 x 10^{56}

Use millimoles Fe^{3+} and Fe^{2+} since these volumes cancel, but use molarities for others:

$[(5.00)^6(0.0167)^2]/([x]^6(x/600)(0.383)^{14}]$ = 7 x 10^{56}
x = $1._3$ x 10^{-7} mmol Fe^{2+}
E = 0.771 − 0.059 log ($1._3$ x 10^{-7}/5.00) = 1.218 V

60 mL: mmol Fe^{3+} = 5.00 − x ≈ 5.00
mmol Fe^{2+} = x
mmol Cr^{3+} = 1.67 − 1/3x ≈ 1.67; M = 0.0152 mmol/mL
mmol $Cr_2O_7^{2-}$ = 0.167 + 1/6x ≈ 0.167; M = 0.00152 mmol/mL
mmol H^+ = 38.3 + 14/6 x ≈ 38.4; M = 0.348 mmol/mL
E = 1.33 − (0.059/6) log $[(0.0152)^2/(0.00152)(0.348)^{14}]$ = 1.28 V

10. 10.0 mL: mmol MnO_4^- added = 0.0200 M x 10.0 mL = 0.200 mmol
mmol Fe^{2+} reacted = 5 x 0.200 = 1.00 mmol = mmol Fe^{3+} formed
mmol Fe^{2+} left = 0.100 M x 100 mL − 1.00 mmol = 9.00 mmol Fe^{2+}
E = 0.771 − 0.059 log (9.00)/(1.00) = 0.715 V

50.0 mL: One half the Fe^{2+} is converted to Fe^{3+} (5.00 mmol each)
E = 0.771 V

100 mL: See Example 14.6.
E = 1.35 V

200 mL: mmol Mn^{2+} = 2.00 − 1/5 x ≈ 2.00 mmol
mmol MnO_4^- = 0.0200 M x 100 mL + 1/5 x ≈ 2.00 mmol
$[H^+]$ = 84 mmol/300 mL = 0.28 M
(We would have to calculate x as at 100 mL to use the Fe equation.)
E = 1.51 − (0.059)/(5) log (2.00)/(2.00)(0.28)^8 = 1.46 V

11. This is a symmetrical reaction, so:
$E_{e.p.}$ = $(n_1E^0_1 + n_2E^0_2)/(n_1 + n_2)$ = [(1)(0.771) + (2)(0.154)]/(1 + 2) = 0.360 V

12. $2Fe^{3+}$ + Sn^{2+} = $2Fe^{2+}$ + Sn^{4+}
   2x   x   2c − 2x   c-x

Neglecting x compared to c:
E = $E^0_{Fe3+,Fe2+}$ − (0.059/n_{Fe}) log ($[Fe^{2+}]/[Fe^{3+}]$)
   = $E^0_{Fe3+,Fe2+}$ − (0.059/1) log (2c/2x)   (1)
E = $E^0_{Sn4+,Sn2+}$ − (0.059/n_{Sn}) log ($[Sn^{2+}]/[Sn^{4+}]$
   = $E^0_{Sn4+,Sn2+}$ − (0.059/2) log (x/c)   (2)

From (1):
(1)E = (1)$E^0_{Fe3+,Fe2+}$ − 0.059 log (c/x)   (3)

From (2):
(2)E = (2)$E^0_{Sn4+,Sn2+}$ − 0.059 log (x/c)   (4)

Adding (3) and (4):
(1)E + (2)E = (1)$E^0_{Fe3+,Fe2+}$ + (2)$E^0_{Sn4+,Sn2+}$ − 0.059 log [(c/x)(x/c)]
E = [(1)$E^0_{Fe3+,Fe2+}$ + (2)$E^0_{Sn4+,Sn2+}$]/[(1) + (2)]

13. $5Fe^{2+}$ + MnO_4^- + $8H^+$ = $5Fe^{3+}$ + Mn^{2+} + $4H_2O$
   5x   x   5c − 5x   c-x

Neglecting x compared to c:
E = $E^0_{Fe3+,Fe2+}$ − (0.059/n_{Fe}) log ($[Fe^{2+}]/[Fe^{3+}]$);
n_{Fe}E = $n_{Fe}E^0_{Fe3+,Fe2+}$ − 0.059 log (5x/5c)   (1)

E = $E^0_{MnO4-,Mn2+}$ − (0.059/n_{Mn}) log ($[Mn^{2+}]/[MnO_4^-][H^+]^8$);
n_{Mn}E = $n_{Mn}E^0_{MnO4-,Mn2+}$ − 0.059 log (c/x) + 0.059 log $[H^+]^8$ (2)

Adding (1) and (2):
n_{Fe}E + n_{Mn}E = $n_{Fe}E^0_{Fe3+,Fe2+}$ + $n_{Mn}E^0_{MnO4-,Mn2+}$ − 0.059 log [(5x/5c)(c/x)] +
(8)(0.059) log [H+]

E = $(n_{Fe}E^0_{Fe3+,Fe2+}$ + $n_{Mn}E^0_{MnO4-,Mn2+})/(n_{Fe} + n_{Mn})$ − [(8)(0.059)/$(n_{Fe} + n_{Mn})$] pH
We will use 0.21M H^+ at equilibrium, assuming the second proton is only slightly ionized (it is only about 2%), so pH = 0.68.

Ee.p. = $[(1)(0.771) + (5)(1.51)]/(1 + 5) - [(8)(0.059)/(1 + 5)] \times 0.368 = 1.33$ V

This compares with 1.33 V obtained in the example calculating the equilibrium concentrations.

14. (a) Before reaction:

$E_{cell} = -0.403 - (0.059/2) \log (1/[Cd^{2+}]) + 0.763 + (0.059/2) \log (1/[Zn^{2+}])$

$= -0.403 - (0.059/2) \log (1/0.0100) + 0.763 + (0.059/2) \log (1/0.250) = 0.319$ V

After reaction, $E_{cell} = 0$. The reaction is:

$$Zn + Cd^{2+} = Zn^{2+} + Cd$$
$$ x \phantom{Cd^{2+}} 0.260 - x$$
$$\approx 0.260 \ M$$

Assuming excess metallic zinc, the 0.0100 M Cd^{2+} has produced 0.0100 M Zn^{2+} and the total $[Zn^{2+}]$ is 0.260 M (x is the equilibrium concentration of Cd^{2+}). Since $E_{Cd} = E_{Zn}$, $E_{Cd} = E_{Zn} = -0.763 - (0.059/2) \log (1/0.260) = -0.780$ V

Equilibrium constant:

$-0.403 - (0.059/2) \log (1/Cd^{2+}) = -0.763 - (0.059/2) \log (1/[Zn^{2+}])$

$-\log K_{eq} = -\log ([Zn^{2+}]/Cd^{2+}]) = (-0.360)/(0.059/2)$

$K_{eq} = 1.6 \times 10^{12}$

(b) Before titration:

$E_{cell} = 0.5355 - (0.059/2) \log [I^-]^3/[I_3^-] + 0.126 + (0.059/2) \log (1/[Pb^{2+}])$

$= 0.5355 - (0.059/2) \log [(1.00)^3/(0.100)] + 0.126 + (0.059/2) \log (1/(0.0100)$

$= 0.691$ V

After reaction:

$$Pb + I_3^- = Pb^{2+} + 3I^-$$
$$ x 0.110 - x \phantom{Pb^{2+}} 1.30 - 3x$$
$$\approx 0.110 \ M \approx 1.30 \ M$$

$E_{Pb} = E_I = -0.126 - (0.059/2) \log (1/0.110) = -0.154$ V

Equilibrium constant:

$0.5355 - (0.059/2) \log [I^-]^3/[I_3^-] = -0.126 (0.059/2) \log (1/[Pb^{2+}]$

$-\log K_{eq} = -\log ([Pb^{2+}][I^-]^3/[I_3^-]) = (-0.126 - 0.5355)/0.059/2)$

$K_{eq} = 2.6 \times 10^{22}$

15. mmol $S_2O_3^{2-} = 6 \times$ mmol $K_2Cr_2O_7$ (Equations 14.9 and 14.10)

$\therefore M \times 1$ mL $= [(0.0490$ mg$)/(294$ mg/mmol$)] \times 6$

$S_2O_3^{2-} = 0.00100 \ M$

$S_2O_3^{2-} + 4I^- + 6H^+ = Se + 2I_2 + 3H_2O$

$\therefore$ mmol $S_2O_3^{2-} = \frac{1}{2}$ mmol $I_2 = \frac{1}{4}$ mmol $S_2O_3^{2-}$

mg Se $= 0.00100 \ M \times 4.5$ mL $\times \frac{1}{4} \times 79.0$ mg/mmol $= 0.089$ mg $= 89 \ \mu g$

$89 \ \mu g/10.0$ g $= 8.9 \ \mu g/g = 8.9$ ppm Se

16. $5H_2C_2O_4 + 2\ MnO_4^- + 6H^+ = 10CO_2 + 2Mn^{2+} + 8H_2O$
    mmol Ca^{2+} = mmol $C_2O_4^{2-}$ = 5/2 mmol MnO_4^-
    $\therefore$ mmol Ca^{2+} = 5/2 x 0.00100 M x 4.94 mL = 0.0124 mmol
    meq Ca^{2+} = 2 x mmol Ca^{2+} = 2 x 0.0124 = 0.0248 meq
    $[Ca^{2+}]$ = 0.0248 meq/5.00 mL = 4.96 meq/L

17. The first titration measures only the As(III) (Na_2HAsO_3).
    $H_2AsO_3^- + I_2 + H_2O = HAsO_4^{2-} + 2I^- + 3H^+$
    mmol As(III) = mmol I_2
    % Na_2HAsO_3 = [(0.150 M x 11.3 mL x 170 mg/mmol)/(2500 mg)] x 100% = 11.5%

    The second titration measures total arsenic: the As(III) that was converted to As(V) in the first titration, plus the As(V) initially present.
    $H_3AsO_4 + 2I^- + 2H^+ = H_2AsO_3 + I_2 + H_2O$
    mmol As = mmol I_2 = ½ mmol $Na_2S_2O_3$
    mmol total As = ½ x 0.120 M x 14.2 mL = 2.47 mmol
    mmol As(III) = 0.150 M x 11.3 mL = 1.70 mmol
    mmol As(V) = 2.47 – 1.70 = 0.77 mmol
    mmol As_2O_5 = ½ mmol As(V)
    $\therefore$ % As_2O_5 = [(0.77 mmol x ½ x 230 mg/mmol)/(2500 mg)] x 100% = 3.5_4 %

18. M_{KMnO4} = [125 mg Fe x 1/5 (mmol MnO_4^-/mmol Fe)]/(55.8 mg/mmol x 1.00 mL $KMnO_4$) = 0.448 M
    M_{tetrox} = [0.448 M_{KMnO4} (mmol/mL) x 0.175 mL_{KMnO4} x 5/4 (mmol tetrox/mmol $KMnO_4$)]/(1.00 mL_{tetrox}) = 0.0980 M
    0.200 M_{NaOH} (mmol/mL) x mL_{NaOH} = 0.0980 M_{tetrox} (mmol/mL) x 1.00 mL_{tetrox} x 3 (mmol NaOH/mmol tetrox)
    mL_{NaOH} = 1.47 mL

19. The concentrations of standards in the diluted samples are 1.2 x 10^{-3} M and 2.4 x 10^{-3} M, respectively. Let x represent the concentration of sulfide in the diluted sample.

    -216.4 = k – 29.6 log x   (1)
    -224.0 = k – 29.6 log (x + 1.2 x 10^{-3})   (2)

    (1) – (2):
    7.6 = 29.6 log [(x + 1.2 x 10^{-3})/x]
    (x + 1.2 x 10^{-3})/x = 1.81
    x = 1.5 x 10^{-3} M
    The sample was diluted 2.5:1. Therefore, the sample concentration is:
    (1.5 x 10^{-3}) x 2.5 = 3.8 x 10^{-3} M sulfide

20. From K_a,
    $[H^+] = K_a[HA]/[A^-]$

During the titration,

$[HA] = (M_{HA}V_{HA} - M_BV_B)/V_{HA} + V_B)$

$[A-] = M_BV_B/V_{HA} + V_B)$

So

$[H^+] = K_a(M_{HA}V_{HA} - M_BV_B)/M_BV_B$

At the equivalence point

$M_{HA}V_{HA} = M_BV_{eq\ pt}$

So $[H^+] = K_a(M_BV_{eq\ pt} - M_BV_B)/M_BV_B$

or

$V_B[H^+] = K_a(V_{eq\ pt} - V_B) = V_B10^{-pH}$ (since $[H^+] = 10^{-pH}$)

At the equivalence point, $V_B = V_{eq\ pt}$, and so the equivalence point corresponds to the zero intercept in the y-axis in a plot of V_B10^{-pH} vs. V_B. The slope is $-K_a$.

21. See the text website for spreadsheet calculation of titration curve and the plot.

CHAPTER 15 VOLTAMMETRY AND ELECTROCHEMICAL SENSORS

1. Back emf = voltage exerted by a galvanic cell which opposes the external applied voltage required for electrolysis to occur.

   Overpotential = the electrode potential in excess of the reversible Nernst potential required for electrolysis to occur.

   IR drop = the voltage drop in a electrolysis cell due to the resistance of the cell and is equal to the product of the cell current and the cell resistance. An increased voltage, equal to the iR drop, is required to maintain the electrolysis.

2. Half-wave potential = the potential of a voltammetric wave at which the current is one- half the limiting current. Depolarizer = a substance that is reduced or oxidized at a microelectrode.

   D.M.E. = dropping mercury electrode. This is the microelectrode used for polarography.

   Residual current = the background voltammetric current, caused by impurities and the charging current. This limits the sensitivity of these techniques.

   Voltammetry = all current-voltage methods using microelectrodes.

3. The supporting electrolyte minimizes the migration current and the iR drop.

4. Electrolytically reduce the Fe^{3+} to Fe^{2+} at 0 V vs. S.C.E. at a large cathode before running the polarogram, using either a platinum gauze or a mercury pool cathode. The Fe^{2+} will not be reduced at –0.4 V and will not interfere with the polarographic lead wave.

5. The complexed metal is stabilized against reduction, and a more negative potential will be required to reduce it, i.e., the half-wave potential is shifted to a more negative potential.

6. A chemically modified electrode consists of an electrochemical transducer and a chemically selective (or catalytic) layer.

7. The electrocatalyst reduces the needed applied potential to generate an amperometric current from an electrochemically irreversible analyte. It does so by reacting with the analyte to be converted to a form which is electrolyzed at the applied potential.

8. The dimensions of an ultramicroelectrode are smaller than the diffusion layer thickness, making the mass transport independent of flow and enhancing the signal-to- noise ratio.

9. Concentration of lead added = C; x = concentration in unknown solution.
   C x 11.0 mL = 1.00 x 10^{-3} M x 1.00 mL
   C = 9.1 x 10^{-5} M

Current increase caused by added lead $= (12.2 - 5.6)$ μ A $= 6.6$ μ A

$(6.6$ μ A$)/(9.1 \times 10^{-5}$ $M) = (5.6$ μ A$)/x$

$x = 7.7 \times 10^{-5}$ M lead

10. Since a current is recorded at zero V, Fe^{3+} is present. Since the wave at -1.5 V exceeds twice the current at zero V, Fe^{2+} is also present. The Fe^{3+} contribution to the wave at -1.5 V is 2 x 12.5 = 25.0 μ A. Therefore, the Fe^{2+} contribution is $30.0 - 25.0 = 5.0$ μ A. The relative concentrations are $[Fe^{3+}]/[Fe^{2+}] = 25.0/5.0 = 5:1$.

11. PFQ

The correct answer here is c. For a reduction reaction at constant current, the potential continues become more and more negative as the oxidized species (A in this case) is converted to the reduced species (B in this case). The potential is given by $E = k - (0.059/n)\log[B]/[A]$, and the log term becomes larger as A is converted to B. If all A is electrolyzed or is decreased to a concentration that can't carry the constant current, the potential will shift to the next reducible substance, e.g., water.

CHAPTER 16   SPECTROCHEMICAL METHODS

1. In the far infrared region, quantized rotational energy transitions occur in absorption. In the mid-infrared region, these are superimposed on quantized vibrational transitions. In the visible and ultraviolet regions, quantized electronic transitions occur in addition to the rotational and vibrational transitions.

2. Paired nonbonding outershell electrons (n electrons) and pi (π) electrons in double or triple bonds.

3. Transitions to a π^* antibonding orbital (n $\rightarrow$ π^* and $\pi \rightarrow \pi^*$ transitions). At wavelengths less than 200 nm, n $\rightarrow$ σ^* transitions may also occur. $\pi \rightarrow \pi^*$ transitions are the most intense ($\varepsilon = 1{,}000 - 100{,}000$ vs. $1{,}000$ for n $\rightarrow \pi^*$).
    Examples are: $\pi \rightarrow \pi^*$ and n $\rightarrow \pi^*$ - ketones
    n $\rightarrow \pi^*$ - ethers, disulfides, alkyl halides, etc. at < 200 nm.

4. There must be a change in the dipole moment of the molecule.

5. Stretching and bending.

6. Absorption in the near-IR region ($0.75 - 2.5$ mm) is the result of vibrational overtones which are weak and featureless. They are due mainly to C-H, O-H, and N-H band stretching and bending motions. NIR is useful for analyzing "neat", i.e., undiluted, samples, and as such, is useful for non-destructive analysis.

7. chromophore - an absorbing group
    auxochrome – enhances absorption by a chromophore or shifts its wavelength of absorption
    bathochromic shift - λ_{max} shifted to longer wavelength
    hypsochromic shift - λ_{max} shifted to shorter wavelength
    hyperchomism – absorption intensity increased
    hypochomism – absorption intensity decreased

8. (a) $CH_2=CHCO_2H$
    (b) $CH_3C=C-C=CCH_3$
    (c)

9. (a) Bathochromic shift, increased absorption

    (b) Bathochromic shift, increased absorption

    (c) No shift, but increased absorption

10. They are highly conjugated, and in alkaline solution loss of a proton affects the electron distribution and hence the wavelength of absorption.

11. Excitation of the metal ion or of the ligand, or via a charge transfer transition (movement of electrons between the metal ion and ligand).

12. absorption = fraction of light absorbed = $1 - (P/P_0)$
    absorbance = $-\log (P/P_0)$ and is proportional to the concentration
    percent transmittance = percent of light transmitted = $(P/P_0) \times 100$
    transmittance = fraction of light transmitted = (P/P_0)

13. Absorptivity is the proportionality constant, a, in Beer's law, when the concentratio units are expressed in g/L. Molar absorptivity, ε, is the proportionality constant when the concentration is expressed in mol/L and is equal to a x f.w.

14. At an absorption maximum, the average absorptivity of the band of wavelengths passed remains more nearly constant as the concentration is changed. The steepness of an absorption shoulder increases as the concentration increases, with the result that the average absoptivity of the wavelengths passed may change.

15. Ultraviolet region: the solvents listed in Table 16.3.
    Visible region: any colorless solvent, including those listed in Table 16.3.
    Infrared region: carbon tetrachloride or carbon disulfide to cover the region of $2.5 - 15 \ \mu$ m.

16. An isosbestic point is the wavelength at which the absorptivities of two species in equilibrium with each other are equal, i.e., where their absorption spectra overlap. The absorbance at this wavelength remains constant as the equilibrium is shifted, e.g., by varying pH.

17. Deviations from Beer's law can be caused by chemical equilibria in which the equilibria are concentration-dependent. These can usually be minimized by suitable buffering. Other deviations can be caused by instrumental limitations, particularly by the fact that a band of wavelengths is passed by the instrument, rather than monochromatic light. These are apparent deviations, when the concentration is so high that the index of refraction of the solution is changed, or when the index of refraction of the sample solvent is different from that of the reference solvent.

18.
Region	Source	Detector
Ultraviolet	H_2 or D_2 dischage tube	Phototube or photomultiplier tube
Visible	Tungsten bulb	Phototube or photomultiplier tube
Infrared	Nernst glower or Gobar	Thermocouple, balometer, or thermister

19. As the slit width is increased, the resolution is decreased, since a wider range of wavelengths is passed. This results in a narrower concentration range over which Beer's law is obeyed because the light passed is less monochromatic. The spectral slit width is the range of wavelengths passed by the slit, whereas the slit width is the physical dimension (width) of the mechanical slit.

20. In a single beam spectrophotometer, the light from the source passes through the sample and falls on the detector, which directly measures the amount of light absorbed. In a double beam instrument, the light from the source is split, with one path going through the sample, and the other around it. The detector measures each beam alternately and the difference in their intensities is read. This minimizes instability due to fluctuations in the power source, amplifier electronics, etc. It also allows for automatic blank correction then the blank solution is placed in the reference path.

21. Near-IR instruments are reliable because of the more intense radiation sources, high radiation throughput, and sensitive detectors, which results in high signal-to-noise ratios.

22. The radiation is dispersed after passing through the sample, and the dispersed radiation falls on the face of a linear diode array. Each diode acts as an individual detector for the different wavelengths, providing measurement of all wavelengths simultaneously.

23. The radiation source is split into two beams, one reflected by a fixed mirror and the other by a moving mirror at 90° from the other. The two reflected beams combine, resulting in an interference pattern for each wavelength, and pass through the sample. The recorded time domain spectrum is an interferogram. Fourier transformation of this gives a conventional frequency domain spectrum. An inerferometer offers advantages of increased throughput and the multiplex advantage of measuring all wavelengths simultaneously. The throughput advantage is particularly valuable in the infrared region where source intensities are weak.

24. The acid form absorbs at 580 nm and the base form absorbs at 450 nm. Referring to Table 16.1, the former would be blue and the latter would be yellow or yellow-green. Filters of the complement colors would be used for each since they must pass only the radiation to be absorbed by each form of the compound, i.e., a yellow filter for the acid form and a blue or violet filter for the base form.

25. Light of high energy (e.g., UV) is absorbed and raises the molecule to a higher electronic energy level. Part of the absorbed energy is lost via collisional processes, whereby the electrons are dropped to the lowest vibrational energy of the first excited state. The electrons return to the ground state from this level, emitting a photon of lower energy and longer wavelength than the absorbed light. Fluorescence is more sensitive than absorption analysis, because the difference between zero and a finite signal is measured, as opposed to the difference between two finite signals in absorption measurements. The sensitivity is then governed by the intensity of the source and the sensitivity and stability of the detector.

26. In dilute solution, when less than about 8% of the light is absorbed (abc < 0.01).

27. An intense UV light source at right angles to the detector, a primary filter or monochromator between the source and the sample to pass the excitation wavelength (i.e., reject the fluorescent wavelength), a secondary filter or monochromator between the sample and the detector to pass only the fluorescent wavelength, and a photocell or photomultiplier detector.

28. Measure its degree of fluorescence quenching (an indirect method).

29. $(2500 \ A^0)(10^{-8} \ cm \ A^{0 \ -1})(10^4 \ \mu m \ cm^{-1}) = 0.25 \ \mu m$

    $(2500 \ A^0)(0.1 \ nm \ A^{0 \ -1}) = 250 \ nm$

30. $\nu = (c/\lambda) = (3.0 \ x \ 10^{10} \ cm \ s^{-1})/[(40 \ nm \ cycle^{-1})(10^{-7} \ cm \ nm^1)]$

    $= 7.5 \ x \ 10^{14} \ cycles \ s^{-1} \ (Hz)$

    $\bar{\nu} = 1/\lambda = 1/[(400 \ nm)(10^{-7} \ cm \ nm^{-1})] = 25,000 \ cm^{-1}$

31. $(2 \ \mu m)(10^{-4} \ cm \ \mu m^{-1}) = 2 \ x \ 10^{-4} \ cm$

    $(2 \ x \ 10^{-4} \ cm)(10^8 \ A^0 \ cm^{-1}) = 20,000 \ \text{Å}$

    $\bar{\nu} = 1/(2 \ x \ 10^{-4} \ cm = 5,000 \ cm^{-1}$

    $(15 \ \mu m)(10^{-4} \ cm \ \mu m^{-1}) = 1.5 \ x \ 10^{-3} \ cm$

    $(1.5 \ x \ 10-3 \ cm)(10^8 \ A^0 \ cm^{-1}) = 150,000 \ \text{Å}$

    $\bar{\nu} = 1/(1.5 \ x \ 10^{-3} \ cm) = \text{Å}$

32. Energy of a single photon $= E = h\nu$

    $\nu = (3.0 \ x \ 10^{10} \ cm \ s^{-1})/[(300 \ nm)(10^{-7} \ cm \ nm^{-1})] = 1.0 \ x \ 10^{15} \ s^{-1}$

    $E = (6.6 \ x \ 10^{-34} \ J\text{-}s)(1.0 \ x \ 10^{15} \ s^{-1}) = 6.6 \ x \ 10^{-19} \ Joule \ per \ photon$

    $(6.6 \ x \ 10^{-19} \ J \ photon^{-1})(6.0 \ x \ 10^{23} \ photons \ einstein^{-1}) = 4.0 \ x \ 10^5 \ J \ einstein^{-1}$

    $(4.0 \ x \ 10^5 \ J \ einstein^{-1})/(4.186 \ J \ cal^{-1}) = 9.5 \ x \ 10^4 \ cal \ einstein^{-1}$

33. At 20% T, $A = -\log 0.20 = 0.70$

    At 80% T, $A = -\log 0.80 = 0.10$

    At 0.25 A, $T = 10^{-0.25} = 10^{-1} \ x \ 10^{.75} = 0.56$

    At 1.00 A, $T = 10^{-1.00} = 0.10$

34. $A = abc$

    $0.80 = a(2 \ cm)(0.020 \ g/L)$

    $a = 20$

35. $M = [(15.0 \ \mu g/mL)/(280 \ \mu g/ \mu mole)] \ x \ 10^{-3} \ mmol/ \mu mole = 5.36 \ x \ 10^{-5} \ mmol/mL$

    $T = 0.350$

    $A = \varepsilon bc = \log(1/T)$

    $\varepsilon \ x \ 2 \ cm \ x \ 5.36 \ x \ 10^{-5} \ mmol/mL = \log(1/0.350) = 0.456$

    $\varepsilon = 4.25 \ x \ 10^3 \ cm^{-1} \ mol^{-1} \ L$

36. (a) At $2.00 \ x \ 10^{-5} \ M$, $T = 1 - 0.315 = 0.685$

    $A = \log 1/0.685 = 0.164$

    At $6.00 \ x \ 10^{-5} \ M$, A it tripled:

    $A = 3.00 \ x \ 0.164 = 0.492$

    (b) $\log 1/T = 0.492$

    $1/T = 3.104$

    $T = -0.322; \ \%T = 32.2$

    % absorbed $= 100.0 - 32.2 = 67.8\%$

37. $\varepsilon = a \ x \ f.w.$

    $286 \ cm^{-1}g^{-1} \ L \ x \ 180 \ g \ mol^{-1} = 5.15 \ x \ 10^4 \ cm^{-1} \ mol^{-1} \ L$

38. $A = \varepsilon bc$; f.w. aniline = 93.1
ε = a x f.w. = (134 $cm^{-1}g^{-1}L$)(93.1 g mol^{-1}) = 1.25 x 10^4 cm^{-1} mol^{-1} L
A = (1.25 x 10^4 cm^{-1} mol^{-1} L)(1.00 cm)(1.00 x 10^{-4} mol L^{-1}) = 1.25

39. $A = \varepsilon bc$
0.687 = 703 x 1 x c
c = 9.77 x 10^{-4} mol/L
9.77 x 10^{-4} mol/L x 279 g.mol x 2 L = 0.528 g tolbutamine

40. $A = \varepsilon bc$; f.w. aniline = 93.1
0.424 = 1.25 x 10^4 x 1 x c
c = 3.40 x 10^{-5} mol/L
3.40 x 10^{-5} mol/L x 93.1 g/mol = 3.17 x 10-3 g/L

But the original solution was diluted 250:25 and 100:10 before measurement.
$\therefore$ total g aniline = 3.17 x 10^{-3} g/L x (250/25) x (100/10) x 0.500 L = 0.158 g
% aniline = (0.158 g/0.200 g) x 100% = 79.0%

41. (a) From the calibration data, Beer's law is obeyed. Hence, the concentration in the unknown
solution can be determined from either a calibration graph or from calculation. Since
standards are treated the same, the urine concentration is in the same units.
(3.00)(0.625/0.615) = 3.05 ppm in the urine, or 0.00305 g/L
(0.00305 g/L)(1.270 L/day) = 0.00387 g/day (Appendix F answer should be divided by 50.)

(b) (0.00305 g/L)/(0.03097 g/mmol) = 0.0985 mmol/L (Also 1/50th for Appendix F answer.)

(c) pH = pK_2 + log ([HPO_4^{2-}]/[$H_2PO_4^-$])
6.5 = 7.1 + log ([HPO_4^{2-}]/[$H_2PO_4^-$])
([HPO_4^{2-}]/[$H_2PO_4^-$]) = $10^{-0.6}$ = 10^{-1} x $10^{0.4}$ = 2.5 x 10^{-1}

42. The concentration of iron(II) in the stock standard solution is
(0.0702 g $Fe(NH_4)_2SO_4.6H_2O$/L[(55.85 g Fe/mol)/(392.1 g $Fe(NH_4)_2SO_4.6H_2O$/mol)]
= 0.0100 g/L = 10.0 mg/L = 10.0 ppm Fe

The corresponding working standards are:
Solution 1   (10.0 ppm)(1.00/100) = 0.100 ppm
Solution 2   (10.0 ppm)(2.00/100) = 0.200 ppm
Solution 3   (10.0 ppm)(10.0/100) = 1.00 ppm
A plot of A vs. ppm gives a straight line drawn through the origin. From this calibration
graph and the absorbance of the sample, the sample concentration is 0.540 ppm.
(0.540 mg/L)(0.100 L) = 0.054 mg Fe in sample.

43. The concentration of nitrate N in the standard solution is
    $(0.722 \text{ g } KNO_3/L)[(14.07 \text{ g N/mol})/(101.1 \text{ g } KNO_3/mol)] = 0.100 \text{ g/L}$
    $= 100 \text{ mg/L} = 100 \text{ ppm N}$
    Hence, the amount of standard added to spiked sample is 0.100 mg, corresponding to an increase in concentration of $0.100 \text{ mg}/100 \text{ mL} = 1.00 \text{ mg/L} = 1.00 \text{ ppm N}$
    (The volume change can be ignored since the solutions were evaporated and ultimately diluted to a fixed volume of 60 mL.

    The net absorbances are
    Sample: $0.270 - 0.032 = -0.238$
    Sample plus standard: $0.854 - 0.032 = 0.822$
    The change in absorbance due to a sample concentration change of 1.00 ppm N is
    $0.822 - 0.235 = 0.587$

    Hence, the concentration of nitrate N in the sample is
    $(1.00 \text{ ppm})(0.238/0.587) = 0.405 \text{ ppm N}$

44. The initial concentrations of A and B are 0.00500 *M*.

    $$AB \quad = \quad A \quad + \quad B$$
    $$0.00500 - x \qquad x \qquad x$$
    $$(x^2)/(5.00 \times 10^{-3} - x) = 6.00 \times 10^{-4}$$

    Solution of the quadratic equation for x yields:
    $x = 1.46 \times 10^{-3} \, M$

    Therefore, the concentration of the complex is $(5.00 - 1.46) \times 10^{-3} = 3.54 \times 10^{-3} \, M$.
    $A = 450 \times 1.00 \times 3.54 \times 10^{-3} = 1.59_3$

45. The concentration of A can be calculated from the absorbance at 267 nm since B does not absorb there:
    $0.726 = (157 \text{ cm}^{-1} \text{ g}^{-1} \text{ L})(1.00 \text{ cm}) c_A$
    $c_A = 4.62 \times 10^{-3} \text{ g/L} = 4.62 \text{ mg/L}$
    $A_{312} = A_{A,312} + A_{B,312}$
    $0.544 = (12.6)(1.00)(4.62 \times 10^{-3}) + (186)(1.00)c_B$
    $c_B = 2.61 \times 10^{-3} \text{ g/L} = 2.61 \text{ mg/L}$

46. $0.805 = k_{415,Ti} \times 1.00 \times 10^{-3}; \; K_{415,Ti} = 805$
    $0.465 = k_{455,Ti} + 1.00 \times 10^{-3}; \; k_{455,Ti} = 465$
    $0.600 = k_{455,V} \times 1.00 \times 10^{-2}; \; k_{455,V} = 60.0$
    $0.685 = 805 \times C_{Ti} + 40.0 \, C_V$
    $0.513 = 465 \times C_{Ti} + 60.0 \, C_V$
    $C_{Ti} = 6.39 \times 10^{-4} \, M, \, C_V = 3.18 \times 10^{-3} \, M$
    $6.39 \times 10^{-4} \, M \times 100 \text{ mL} = 0.0693 \text{ mmol Ti}$
    $0.0693 \text{ mmol} \times 47.9 \text{ mg/mmol} = 3.32 \text{ mg Ti}$

%Ti = (3.32 mg/1000 mg) x 100% = 0.332%
3.18 x 10^{-3} M x100 mL = 0.318 mmol V
%V = (16.2 mg/1000mg) x 100% = 1.62%

47. From Beer's law, fraction of energy absorbed = $1 - (P/P_0) = 1 - 10^{-abc}$
$P_0 - P = P_0(1 - 10^{-abc})$ = quantity of light absorbed
F $\propto$ quantity of light absorbed
F $\propto$ $P_0(1 - 10^{-abc})$
F = $\phi P_0(1 - 10^{-abc})$, ϕ = proportionality constant

48. We solve (b) and (c) first since we can use (c) to calculate (a).
We extracted 1,000 mL of sample to eventually get 5 mL of eluite from the cartridge.

(b) P_{ideal} is therefore = 1000/5 = 200

(c) During spiking we took 5 x10^{-8} mol Pb additional Pb that went into 5 mL, this would
represent a concentration increase (ΔC) of (5 x 10^{-8} mol/5mL) x 10^3 mL/L = 1 x 10^{-5} M.
Ideally this would have led to an absorbance increase (ΔA) which are related by Beer's
law as $\Delta A = \varepsilon b \Delta C$, the theoretical value of ΔA would thus be

$$2.0 \text{ x } 10^4 \text{ L mol}^{-1} \text{ cm}^{-1} \text{ x } 1.0 \text{ cm x } 1 \text{ x } 10^{-5} \text{ mol L}^{-1} = 0.20$$

We observe ΔA = 0.44-0.32 = 0.12; the efficiency is 0.12/0.2 *100% = 60% of
theoretical.

$$P_{expt} = 0.6 \text{ x } P_{ideal} = 0.6 * 200 = 120$$

(a) Observed absorbance for sample is 0.32. from Beer's Law, the concentration in the final
solution is

$$= a/(\varepsilon b) = 0.32/(2.0 \text{ x } 10^4 \text{ x } 1) = 1.6 \text{ x } 10^{-5} \text{ M}$$

Accounting for a preconcentration factor of 120, the concentration in the original sample is:
$$1.6 \text{ x } 10^{-5} \text{ M}/120 = 1.3 \text{ x } 10^{-7} \text{ M}$$

Alternatively, the concentration is simply 0.32/0.12 times that of the spike in the original
sample.

49. $\lambda = 2tn/n$
For n =1, $\lambda = 2*200*1.377/1$ nm = 551 nm
For n =2, $\lambda = 2*200*1.377/2$ nm = 225 nm

Higher order constructive interferences will occur at wavelengths lower still. A white LED
has no emission below ~400 nm. Therefore, only n= 1 is relevant and the light transmitted is
at 551 nm, where the source actually has its maximum emission.

50. (a) Assume no crosstalk between the top and the bottom half, with $0.5\,P_0$ incident on each half. Light transmitted through the top half therefore is:

Light transmitted through top half $= 0.5\,P_0\,(10^{-\varepsilon b_1 c})$
Light transmitted through bottom half $= 0.5\,P_0\,(10^{-\varepsilon b_2 c})$
Total light transmitted $= 0.5\,P_0\,(10^{-\varepsilon b_1 c} + 10^{-\varepsilon b_2 c})$
$\quad$ Overall transmittance $=$ Total light transmitted$/P_0 = 0.5\,(10^{-\varepsilon b_1 c} + 10^{-\varepsilon b_2 c})$
$\quad$ Overall absorbance $= -\log\,(\text{transmittance}) = -\log 0.5 - \log\,(10^{-\varepsilon b_1 c} + 10^{-\varepsilon b_2 c})$
$\quad = \log 2 - \log\,(10^{-\varepsilon b1c} + 10^{-\varepsilon b2c})$

(b) At low concentrations, the first term within the parentheses will be close to 1 and small variations in c will have little effect on it. In contrast, at large values of c, the second term will be essentially zero and small variations in c will have little effect on it. Consequently in a A vs c plot, the slope will be εb_2 as $c \to 0$ and the slope will be εb_1 as $c \to \infty$.

51. (a) 256 nm. The other peak is below the ethanol solvent cut off of 220 nm.

(b) Deuterium or xenon-arc. LEDs are also available at this wavelength.

(c) Quartz or fused silica cells.

52. PFE. Fourier transformation of a square wave
Consider the simple square wave, $S(t) = +1$ from $t = 0$ to $t = T/2$ and $S(t) = -1$ from $t = T/2$ to $t = T$.

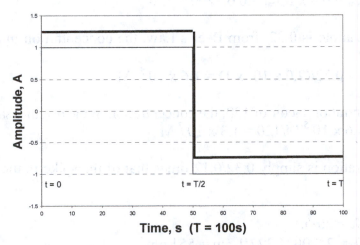

To transform this time signal into its frequency components, we need to get the a_n and b_n amplitude factors. This is done by performing the integrals above. Our example is particularly easy to integrate, once we realize that each integral can be broken into two parts, the first is an integral from 0 to T/2, and the other is from T/2 to T. For each part, then, the function S is just a constant (either 1 or -1). The frequencies are $\nu_1 = 1/100$, $\nu_2 = 2/100$, $\nu_3 = 3/100$, etc.

We would not solve the integrals for a_n and b_n, but familiarity with integral calculus will allow you to easily derive:

For all even values of n, $a_n = 0$
For all odd values of n, $a_n = 4/(\pi n)$
For all values of n, $b_n = 0$

Under these conditions, the Fourier sum :
$$S(t) = \sum (a_n \sin(2\pi v_n t))$$

In the spreadsheet Fourier transform.xlsx we calculate the amplitude at 1 sec intervals for the first 29 odd values of v (you can choose as many as you wish); (i.e., $A_1(t) = a_1 \sin(2\pi * 0.01 * t)$, $A_3(t) = a_3 \sin(2\pi * 0.03 * t)$, and $A_5(t) = a_5 \sin(2\pi * 0.01 * t)$, etc.); we do the calculation for T = 0–100 sec.

We plot the the waves for the first three values of v and their sum in one chart. In another chart we plot the sum of first three, first seven and first fifteen. You will note that using more of the terms, the summed function gets closer to representing the original signal. That is, the component waves will add together to give the original function.

Finally, we plot the spectrum of the waves; this is the Fourier transform of the original time signal. The intensity of a wave is simply the square of its amplitude, the amplitude here is the coefficient a_n. So we simply plot a_n^2 vs v.

See the text website for an Excel spectrum calculation.

53. PFE, four component mixture. See the text website for an Excel solution using the "jackknife" approach.

CHAPTER 17   ATOMIC SPECTROMETRIC METHODS

1. Only a small fraction of atoms in a flame are termally excited, 0.01% or less.

2. A solution is aspirated into a flame where atomic vapor of the elemental constituents (metals) is produced. A certain small fraction of the atoms is thermally excited to higher electronic levels, and when these return to the ground state, photons of characteristic wavelength are emitted. These emitted photons are measured in flame emission spectrophotometry. In atomic absorption spectrophotometry, the absorption of the characteristic wavelengths of light by the ground state atoms is measured.

3. Flame emission spectrometry requires a flame-aspirator for atomization and excitation of the sample, a monochromator for separating the emitted wavelengths, and a detector, usually a photomultiplier . Atomic absorption spectrophotometry is similar, but requires in addition a light source, usually a sharp line source. The two techniques are similar in sensitivity for a large number of elements, with each being more sensitive for a number of specific elements. They are both subject to essentially the same chemical and ionization interferences, but flame emission spectroscopy is generally more subject to spectral interferences.

4. In order to obtain higher sensitivity, greater specificity, and linearity in calibration curves.

5. Because the discrete electronic transitions are not superimposed on rotational and vibrational transitions, since atoms do not undergo these latter two modes of transition.

6. The red zone in a reducing nitrous oxide-acetylene flame results from the presence of highly reducing radical species, such as CN and NH radicals.

7. The atomization efficiency (the fraction of sample converted to atomic vapor) is much higher than in a flame, as is the residence time in the absorption path.

8. If the internal standard and analyte are chemically similar, changes in these populations due to chemical reactions, changes in aspiration rate or flame conditions and so forth will be similar for both elements. Hence, the ratio of their populations will remain constant and so will the ratio of their measured signals.

9. Multiple elements are routinely measured by flame photometry; in yesteryears, with hotter flames, many more elements were in fact essentially simultaneously measured by flame emission, with a scanning monochromator. If it were done today, it will likely be done with a polychromator and an array or a 2-D detector. In principle (also as outlined below), simultaneous atomic fluorescence can be observed by directing different hollow cathode lamps to the same spot in the atom population in an atomizer and observing the fluorescence at right angles, using a single scanning monochromator and a detector or a polychromator and an array detector. In AAS, we go along the long axis of the atom plume and read the attenuation of the same beam on the other side. Putting multiple HCLs along the long axis of a plume will be difficult and on the other side it will be difficult to read the transmitted non-

collinear beams by a single monochromator/detector. If we try to read the different beams by different monochromators/detectors, there will still be a problem locating them spatially. The development of continuum source AAS that utilizes a high intensity continuum sources, a high resolution polychromator and an array detector permits essentially simultaneous multielement measurement.

We can design a multielement AFS instrument the same way as a continuum source AAS, merely reading the light at right angles. However, within the extremely narrow bandwidths that atoms are excited, a continuum source, even a 300 W hot spot xenon lamp, does not provide as much excitation power as a boosted hollow cathode; sensitivity will not be competitive with that obtainable from dedicated boosted HCL's all other parameters being equal.

10. (a) In a hollow cathode lamp or any other line sources, an element is excited not just to the lowest excited state but other states as well, and emission occurs from these states as well with the state following the photon emission not necessarily being the emission state. Thus there are several emission lines in a HCL and the situation is obviously worse in multielement HCls. A monochromator is hence needed. In favorable cases, e.g., using a Na-lamp to produce the D-lines almost exclusively, it is indeed possible to follow the absorption of this line without any monochromator as both Walsh and L'vov did in their scouting experiments.

   (b) Go to the site http://physics.nist.gov/PhysRefData/Handbook/element_name.htm mentioned in Example 17.1 and look up sodium. You can look at the persistent line listings for both the neutral Na atom and the singly ionized Na^+. In the first case you see that the strongest Na emission is at 588.95 nm, while the strongest emission from Na^+ is at 328.56 nm. Not only will more expensive UV sensitive detectors will be required to monitor the latter, estimating the excited state population differences of Na^* vs. Na^{+*} as in Figure 17.1, there will be far fewer Na^{+*} than Na^* at typical Flame temperatures; it will be less sensitive. This will be the same for other alkali metal ions (check!)

11. The cathode is constructed of the element to be determined, or an alloy of it, and the anode is an inert metal such as tungsten. The lamp is filled with an inert gas under reduced pressure. A high voltage across the electrodes causes the gas to ionize at the anode, and these positive ions bombard the cathode, causing the metal to sputter (vaporize) and become electronically excited. These excited atoms emit the characteristic lines of the metal when they return to the ground state.

12. In the premix chamber burner, the fuel and support gases are introduced into a chamber where they are mixed before entering the burner head where they combust. It is limited to relatively low-burning velocity flames, such as air-acetylene. It cannot be used with oxygen-supported flames.

13. To convert it to an a.c. signal which can be selectively measured with an a.c. detector tuned to its modulation frequency and thereby discriminate from d.c. emission from the flame.

14. Probably non-specific background absorption by something in the sample extract. It can be corrected for using background correction techniques described in the text..

15. In order to atomize refractory elements like Al and V, that from stable oxides in cooler flames.

16. To suppress ionization.

17. To minimize ionization of the elements. They have few chemical interferences.

18. The presence of the potassium suppresses ionization of the sodium, causing an increase in free sodium atoms and hence atomic emission.

19. $SnCl_2$ reducing agent produces mercury vapor from inorganic mercury, while $NaBH_4$, a more powerful reducing agent, will also reduce organically bound mercury. The difference in signals from the two procedures is used to calculate the organically bound mercury.

20. The hydride vapor of the analyte element is separated from the sample matrix, reducing matrix effects.

21. A radiofrequency of 5-75 MHz at 1-2 kW power excites argon gas flowing in a quartz tube. The ionized argon conducts and the alternating magnetic field generates an eddy current within the conduction gas, and the dissipated energy heats the plasma to high temperatures up to 10,000 K. The plasma serves as an excellent atomizer and excitation source for many elements.

22. In laser ablation, a high energy UV laser pulse is directed onto a solid sample surface. The deposited energy vaporizes/aerosols the sample spot, and the vapor/aerosol is carried into an ICP for elemental analysis.

23. AFS is best used for hydride-forming elements, and for mercury via cold vapor generation of Hg vapor. Commercial AFS mercury analyzers are available.

24. Sensitivity = concentration needed to give 1% absorption or 0.00044 A (= -log 0.99).
    T = 0.920
    A = log 1/0.920 = log 1.09 = 0.0362
    (0.0044 A/0.0362 A) = (x ppm/12 ppm)
    x = 1.5 ppm per 0.0044 A = sensitivity

25. (0.050 ppm/0.70 ppm) = (0.0044 A/x A); x = 0.062 A
    -log T = 0.062; T = 0.87; % absorbed = 13%

26. $T = 2250^0$ C = 2523 K

    From the term symbols, $J_e = 1$ and $J_0 = 0$
    g= 2J + 1

$g_e/g_0 = [2(1) + 1]/[2(0) + 1] = 3/1$

$\lambda = 228.8$ nm $= 2.288 \times 10^{-5}$ cm

$\nu = c/\lambda = (3.00 \times 10^{10}$ cm g-1$)/(2.288 \times 10^{-5}$ cm$) = 1.31 \times 10^{15}$ s^{-1}

$E_e - E_0 = h\nu = (6.22 \times 10^{-27}$ erg-s$)(1.31 \times 10^{15}$ s$^{-1}) = 8.68 \times 10^{-12}$ erg

$N_e/N_0 = g_e/g_0\, e^{-h\nu/kT}$

$= (3/1) \exp[-(8.68 \times 10^{-12}$ erg$)/(1.380 \times 10^{-16}$ erg K$^{-1})(2523$ K$)]$

$= 3e^{-24.93} = 4.5 \times 10^{-11}$

$\% N_e = [N_e/(N_e + N_0)] \times 100\% = [(4.5 \times 10^{-11})/(1.00)] \times 100\% = 4.5 \times 10^{-9}\%$

27. Stock Ca solution:

1.834 g [(40.08 g/mol Ca)/147.02 g/mol CaCl$_2$.2H$_2$O] = 0.0500 g Ca

500 mg Ca/L = 500 ppm Ca

Therefore, second stock solution = 50.0 ppm

Working standards = 2.50 , 5.00, and 10.0 ppm Ca

Net signals (subtract 1.5 cm for blank):
Standards: 9.1, 18.6, 37.0 cm
Sample: 28.1 cm

Plot of net signal vs. concentration of standards gives straight line through the origin. Sample reading on the graph corresponds to 7.62 ppm Ca. Therefore, concentration in original sample is 7.62 ppm x 25 = 190 ppm Ca

28. The volume of added LiNO$_3$ can be neglected compared to 1 mL (1% change – the instrumental reproducibility is about 2%). Hence, the concentration of Li added to the sample solution is 0.010 M x (0.01/1) = 1.0 x 10^{-4} M. The net increase in signal is due to this added concentration: 14.6 – 6.7 = 7.9 cm. Assuming linearity, a direct proportionality applies:

(7.9 cm)/(1.0 x 10^{-4} M) = (6.7 cm)/(x)

x = 8.5 x 10^{-5} M in the diluted serum

8.5 x 10^{-5} M x (1/0.1) = 8.5 x 10^{-4} M in the undiluted serum

8.5 x 10^{-4} mmol/mL x 6.94 mg/mmol = 5.9 x 10^{-3} mg/mL

= 5.9 μg/mL = 5.9 ppm

29. A direct proportionality exists between concentration and the decrease in the silver absorbance signal from the blank reading.

(12.8 – 5.7 cm)/(100 ppm) = (12.8 – 6.8 cm)/(x)

x = 84 ppm in sample

30. PFP

# of M&M added	0	10	20	30	40
Total mass (g)	52.7	58.1	63.6	68.5	74.4
net mass (without bag)	42.7	48.1	53.6	58.5	64.4

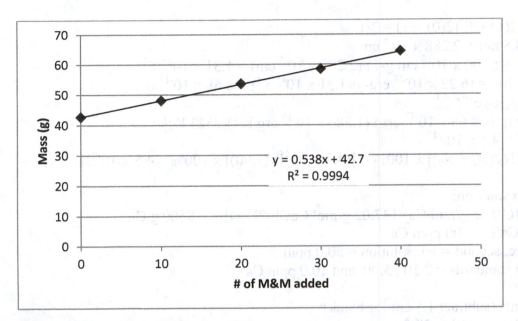

Obtain the intercept at y=0, x=- 79.4

Therefore the original number of M&Ms in the bag is 79.

CHAPTER 18   SAMPLE PREPARATION: SOLVENT AND SOLID-PHASE EXTRACTION

1. The distribution coefficient is the ratio of the concentration of the extracted solute in the nonaqueous phase to its concentration in the aqueous phase. The distribution ratio is the ratio of all forms of the solute in the two phases and accounts for ionization, dimerization, etc.

2. Extract the nitrobenzene from acid solution into benzene. The aniline is ionized ($C_6H_5NH_3^+$) in the acid and will not extract.

3. (1) Ion-association extraction systems: The metal ion is incorporated into a bulky molecule, either charged or uncharged, which associates with another ion of opposite charge to form an ion-pair, or else the metal ion associates with another ion of great size. An example is the coordination of solvent, i.e., the solvent extraction of iron(III) from HCl solutions into ether.

   (2) Metal chelate extraction systems: The metal ion forms an uncharged chelate with an organic chelating agent, which is soluble in the nonaqueous solvent. An example is the solvent extraction of aluminum with 8-hydroxyquinoline into chloroform.

4. The chelating agent, a weak acid, distributes between the two phases, it ionizes in the aqueous phase, the ionized reagent chelates with the metal ion, and the chelate distributes between the two phases.

5. Theoretically, the upper limit is the solubility limit of the chelate in the organic solvent, and at the other extreme, unmeasurable amounts can be extracted.

6. An increase in pH by 1 unit or an increase in reagent concentration by ten-fold will increase the extraction efficiency the same amount.

7. The sample and solvent are heated in a closed vessel under pressure to temperatures above the ambient boiling point of the solvent. The elevated temperature accelerates dissolution of analytes in the solvent.

8. The sample and solvent are heated in a microwave oven in a closed vessel, which accelerates the extraction process.

9. Extractions are achieved using microparticles coated with hydrophobic functional groups. These are the same particles used in high performance liquid chromatography. The hydrophobic layer acts as the extracting solvent. The particles are placed in small cartridges, pipet tips, or on disks. Many samples can be processed simultaneously and automatically. Small volumes of eluting solvent are needed, reducing organic solvent waste.

10. SPME is a solvent-less extraction technique in which analytes are adsorbed onto a micro fused silica fiber coated with a solid adsorbent or an immobilized polymer. Following

adsorption, the analyte is thermally desorbed. DLLME injects a few microliters of a binary solvent mixture, one being miscible with the aqueous sample (e.g., MeOH). The immiscible one is generally heavier than water, the rapid injection producing fine droplets which promotes very rapid extraction. Centrifugation is used to recover the heavier extractant.

11. Equation 18.9:
$$\% \, E = \{([S]_0 V_0)/[S]_0 V_0 + [S]_a V_a)\} \times 100 \tag{1}$$

Also,
$$D = ([S]_0/[S]_a \text{ (assuming no side reactions)} \tag{2}$$

Divide the numerator and denominator of (1) by $[S]_a$
$$\%E = \{([S]_0/[S]_a)V_0)/[([S]_0/[S]_a)V_0 + V_a]\} \times 100 = [(DV_0/DV_0 + V_a)] \times 100$$

Divide top and bottom by V_0:
$$\% \, E = [D/(D + V_a/V_0)] \times 100$$

12. The concentration of benzoic acid present in the dimer is twice the dimer concentration. Hence,
$$D = ([HBz]_e + 2[(HBz)_2]_e)/([HBz]_a + [Bz^-]_a) \tag{1}$$

From Equation 18.6,
$$[HBz]_e = K_D[HBz]_a \tag{2}$$
For the dimer, $[(HBz)_2]_e = K_p[HBz]_e^2 = K_p K_D^2[HBz]_a^2$ (3)

From Equation 18.5,
$$[Bz^-]_a = K_a[HBz]_a/[H^+]_a \tag{4}$$

Substituting (2), (3), and (4) in (1),
$$D = (K_D[HBz]_a + 2K_p K_D^2[HBz]_a^2)/([HBz]_a + K_a[HBz]_a/[H^+]_a)$$
$$D = K_D(1 + 2K_p K_D[HBz]_a)/(1 + K_a/[H^+]_a)$$

This is identical to Equation 18.8 except the term in parentheses in the numerator is added. From this, the distribution ratio and extraction efficiency are now dependent on the concentration of the extracting species, in addition to the pH.

13. $x = \%$ extracted in a single step
$$x + (100 - x)(x/100) = 96$$
$x = 80\%$ extracted in a single extraction
$$80\% = (100D)/(D + 100/50)$$
$$D = 8.0$$

14. $\%E = (100 \times 2.3)/(2.3 + 25.0/10.0) = 48\%$

15. At $V_a = V_0$, $90\% = (100D)/(D + 1)$
    $D = 9$
    $\therefore$ when $V_a = 0.5V_0$,
    $\%E = (100 \times 9)/(9 + 0.5) = 95\%$

16. PFP
    In the aqueous phase, $K_a = [RCOO^-]_w*[H^+]/[RCOOH]_w = 10^{-6.00}$.
    At pH=6.00, $[RCOO^-]_w = [RCOOH]_w$.

    After the two phases reached an equilibrium, $[RCOO^-] + [RCOOH]_w = 0.0080$ M, on the basis of the fact that $[RCOO^-]_o=0$ and the mass balance. Since $[RCOO^-]_w = [RCOOH]_w$ at pH=6.00, $[RCOOH]_w=0.0040$ M.

    The partition coefficient $K = [RCOOH]_o/[RCOOH]_w = 3.0$ (1)
    At pH= 7.00, $[RCOO^-]_w/[RCOOH]_w = 10$ (from pKa=6.00)   (2)
    From mass balance, $[RCOOH]_o+[RCOO^-]_w+[RCOOH]_w=0.020$ M   (3)

    Substitute equations (1) and (2) to (3), we have
    $[RCOOH]_w=1.4_3*10^{-3}$ M.

    From equation (2), $[RCOO^-]_w=1.4_3*10^{-2}$ M.

    The formal concentration of RCOOH in the aqueous solution is
    $[RCOO^-]_w+[RCOOH]_w= 0.016$ M.

17. With $V_0 = V_a = 10$ mL,
    $\%E = (100 \times 25.0)/(25.0 + 1.0) = 96.2\%$ extracted, leaving 3.8%.

    With $V_0 = 0.50V_a = 5.0$ mL,
    $\%E(100 \times 25.0)/(25.0 + 2.0) = 92.6\%$ extracted first time, leaving 7.4%.

    On second extraction,
    $0.074 \times 7.4 = 0.55\%$ remains.
    $\therefore$ two extractions with half the volume of organic solvent is more efficient.

18. 1st extraction: 30% remains
    2nd extration: $0.30 \times 30 = 9.0\%$ remains
    3rd extraction: $0.30 \times 9.0 = 2.7\%$ remains

19. PFP
    For this problem, we derive the standard additions calculation:
    f = final, i = initial, C = concentration, V = volume, P = preconcentration factor
    $C_f = PC_i$ $P = V_i/V_f$ (ideal) $A = \varepsilon bc$

    $C_{std,i}$ = concentration standard, providing mol_{std} in V_i (spike in mol)
    Measurements in V_f, but related to V_i.

Unknown sample with preconcentration:

$$A_{unk} = \varepsilon b C_{f,unk} = \varepsilon b P C_{i,unk} = \varepsilon b P (mol_{unk}/V_i)$$

Spiked unknown sample with preconcentration:

$$A_{spike} = \varepsilon b C_{f,,spike} = \varepsilon b P C_{i,spike} = \varepsilon b P \left(\frac{mol_{unk} + mol_{std}}{V_i}\right)$$

SAM plot:

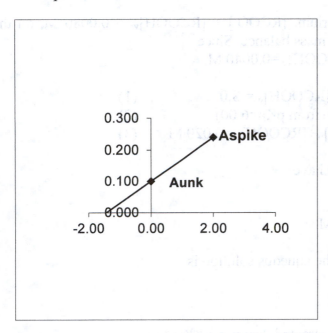

X-axis is concentration of standard prior to preconcentration, $C_{std,i,}$ Y-axis is absorbance. X-intercept is $-C_{i,unk}$.

Slope = $\varepsilon b P$ (get apparent ε if $P_{expt} < P_{ideal}$)

Note: $\varepsilon b P$ = constant, math independent of ε, b, P, so does not matter whether $P_{expt} = P_{ideal}$ to get $C_{i,unk}$.

$$\frac{A_{unk}}{C_{i,unk}} = \varepsilon b P = \frac{A_{spike}}{C_{i,unk} + C_{i,std}}$$

Data: A_{unk}, A_{spike}

Known: $C_{i,,std}$

Rearranging: $C_{i,unk} = C_{i,std}\left(\dfrac{A_{unk}}{A_{spike} - A_{unk}}\right)$

(a)

$$C_{i,unk} = C_{i,std}(\frac{A_{f,unk}}{A_{f,spike} - A_{f,unk}})$$

$$C_{i,std} - \frac{5.0x10^{-8} mol}{1000mL} = 5.0x10^{-8} M$$

$$C_{i,unk} = 5.0x10^{-8} M(\frac{0.32}{0.44 - 0.32}) = 1.3x10^{-7} M$$

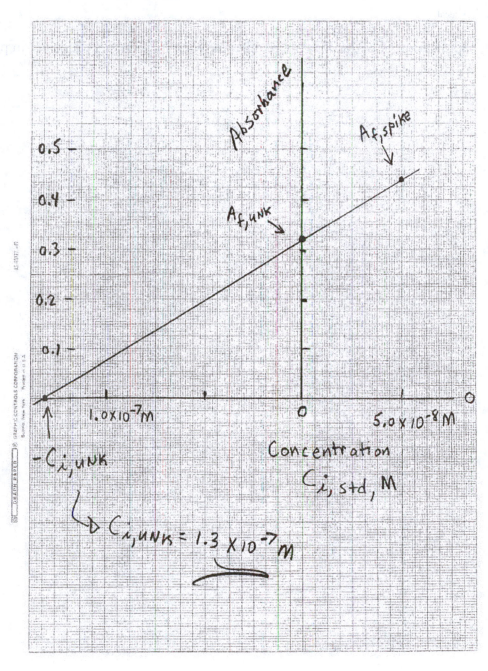

(b)

$$P_{ideal} = \frac{V_i}{V_f} = \frac{1000mL}{5mL} = 200$$

(c)

Using $A_{f,unk} = \varepsilon b P C_{i,unk}$

$$P = \frac{0.32}{(2.0x10^4)(1)(1.3x10^{-7})} = 123 \quad \text{This is } P_{expt.}$$

$P_{exp} < P_{ideal}$

Note, application of SAM does not require knowing P_{expt} since it ratios out in part (a).

CHAPTER 19   CHROMATOGRAPHY: PRINCIPLES AND THEORY

1. Chromatography is a physical method of separation in which the components to be separated are distributed between two phases, one of which is stationary, while the other moves in a definite direction.

2. There is an equilibrium distribution of the solutes between two phases, one mobile and one stationary. The solutes are eluted from the stationary phase by movement of the mobile phase. Due to differences in the distribution equilibria for different solutes, they are eluted at different rates.

3. Chromatographic techniques include adsorption, partition, ion exchange, and size exclusion. Gas chromatography utilizes the first two, while liquid chromatography utilizes all four.

4. The van Deemter equation describes H as a function of velocity of carrier gas or eluent: $H = A + B/\bar{u} + C\bar{u}$. A is the eddy diffusion term, B the molecular diffusion term, and C the mass transfer term.

5. The Golay equation applies to open tubular columns. It does not contain the eddy diffusion (A) term since there is no packing.

6. The Huber and Knox equations contain terms that account for mass transfer in both the stationary and the mobile phase.

7. PFP Increasing K as defined ($= C_s/C_m$) will lead to a greater fraction in the stationary phase And ,therefore, a longer retention time.

8. $N = 16(65/5.5)^2 = 2.2_3 \times 10^3$ plates
   $H = (3 \text{ ft} \times 12 \text{ in/ft} \times 2.54 \text{ cm/in})/(2.2_3 \times 10^3 \text{ plates}) = 0.041 \text{ cm/plate}$

9. When the two peaks are just resolved, the peak base widths will be 15 s, since the difference between their retention times is 15 s. The number of theoretical plates required to elute the last peak is then:
   $N = 16(100/15)^2 = 710$ plates

   The column length is:
   710 plates x 1.5 cm/plate = $1.0_6 \times 10^3$ cm

10. $N_{eff} = 16(t_R/w_b)^2$; $H_{eff} = L/N = 300 \text{ cm/N}$ (use of t_R' gives the *effective* theoretical plates)

mL/min	t_R'	t_R'/w_b	N	H (cm/plate)$_{eff}$
120.2	4.31	12.3	2420	0.123
90.3	4.88	12.5	2500	0.120
71.8	5.43	12.6	2550	0.118
62.7	5.73	12.2	2380	0.126
50.2	6.38	11.8	2230	0.135
39.9	7.25	10.7	1830	0.164
31.7	8.21	10.1	1640	0.183

The optimum flow velocity is near 75 mL/min.

11. Calculate the α-values between the peaks, and from these the number of plates needed to obtain the more difficult one.
$\alpha_{AB} = 1.85/1.40 = 1.32$
$\alpha_{BC} = 2.65/1.85 \quad = 1.43$

Peaks A and B are the poorest resolved, and so a separation factor of 1.32 is needed. The mean value of the two peaks is $(1.40 + 1.85)/2 = 1.62$. From Equation 19.30, the number of plates required for the separation is:

$N_{req} = 16(1.05)^2[1.32/(1.32 + 1)]^2[(1.62 + 1)/1.85]^2 = 602$ plates required.

Hence, compounds A and B won't quite be separated with a resolution of 1.05. The actual resolution of the two peaks is:
$R_s = \frac{1}{4}(500)^{1/2}[(1.32 - 1)/1.32][1.85/(1.62 + 1)] = 0.96$

12. In the first case, the analytes are very highly retained (large k) in a system with good efficiency (high N). Only minimal selectivity would be needed. In the second case, poor resolution might result, despite good selectivity, if retention and efficiency were relatively low. These relationships can be reasoned based on the master resolution equation.

13. See text website for spreadsheet and chart of van Deemter plot.

CHAPTER 20   GAS CHROMATOGRAPHY

1. The stationary phase is a solid or liquid, and the mobile phase is a gas. The solute exists in the vapor state, usually accomplished by heating, and it distributes between the stationary phase and the gas phase.

2. All gases, most non-ionized organic molecules up to C25, volatile derivatives of non- volatile compounds.

3. Gas-solid and gas-liquid are the two types of gas chromatography.

4. Packed columns can have 1,000 plates/m, and a representative 3 m column has 3,000 plates. A capillary column (WCOT) typically has 5,000 plates/m, and 250,000 plates for a 50 m column.

5. Wall coated open-tubular (WCOT) columns have thin film of stationary phase on the capillary wall. Support coated open-tubular (SCOT) columns have solid microparticles coated with stationary phase attached to the wall. They have higher capacity but lower resolution. Porous layer open-tubular (PLOT) columns have solid particles attached to the wall for adsorption chromatography.

6. (a) Thermal conductivity. The difference in the thermal conductivity of the carrier gas and the carrier gas plus solute is measured with a Wheatstone bridge, as the solute is eluted. Solutes generally have a lower thermal conductivity than the carrier gas. A carrier gas with a high thermal conductivity is used, e.g., helium or hydrogen.

    (b) Flame ioinization. The eluted solutes are burned in a hydrogen flame to produce ions, which are collected by a pair of charged electrodes. The resulting current is measured. This detector is insensitive to water, allowing separation of solutes in aqueous solution.

    (c) Electron capture. An electron source (electrical or beta-ray) provides a preselected current to an anode collector. Compounds that capture electrons cause a decrease in this current, which is recorded. A carrier gas with a low excitation energy is used, such as nitrogen or hydrogen, to prevent ionization of compounds.

7. (a) Thermal conductivity. A general detector with relatively low sensitivity.

    (b) Flame ionization. A general detector with high sensitivity, about 1,000 times that of the T.C. detector.

    (c) Electron capture. Specific for a limited number of substances with high electron capture affinity, e.g., chlorine-containing compounds. High sensitivity.

8. A lower initial temperature can be used so faster eluting compounds can be better resolved, and higher temperatures for more strongly retained compounds so they elute faster with less broadening.

9. For fast GC, use a narrow, short column with thin stationary film and a light carrier gas. A fast responding detector is required.

10. 760 torr/14.7 psi = 51.7 torr/psi
    40.0 psig x 51.7 torr/psi = 2070 torr above atmospheric pressure
    2070 torr + 740 torr (atm. press.) = 2810 torr inlet pressure

11. See text website.. The calculated concentration is 20.9 ppm.

CHAPTER 21   LIQUID CHROMATOGRAPHY AND ELECTROPHORESIS

1. In gas chromatography, a carrier gas flows, a low pressure system. In HPLC, a liquid mobile phase must be pumped, requiring high-pressure pumps. Most GC uses capillary columns coated with the stationary phase. HPLC employs different types of particles for the stationary phase. Various gas phase detectors are used in GC. HPLC can employ a variety of detectors for solution phase measurements. In GC, the van Deemter plot is highly dependent on the carrier gas flow rate, see Figure 19.8. But in HPLC, there is less dependence on flow rate, especially for small particle sizes, see Figures 19.9, 21.3 and 21.25.

2. See Figure 21.1 for a block diagram of a basic HPLC system. A pump, injector, column and a detector are minimally needed. An autosampler and a data system are very useful. The ability to degas solvents and generate gradients are highly useful.

3. (a) Modes of separation: Normal phase, reverse phase, ion exchange, ion, hydrophilic interaction, aqueous normal phase, size exclusion, ion exclusion, and affinity chromatography. See Section 21.1 for brief descriptions of each. See HPLC Subclasses.

    (b) Physical nature of column (stationary phases): High-purity silica; perfusion packings (microporous, perfusion, nonporous, and core-shell particles); ion-exchange resins (see also Fig. 21.7 for modern ion-exchangers); monolithic; chiral phases; alumina; zirconia titania, porous graphitic carbon. See Section 21.2 for descriptions of each.

4. Most classical ion exchange resins are made of a polystyrene polymer, crosslinked with divinylbenzene. They are functionalized with strongly or weakly acidic groups, supplied in acid or sodium forms (for cation exchange resins), or strongly or weakly basic groups, supplied in hydroxide or chloride forms (for anion exchange resins). The eluent is an acid to elute cations that have been exchanged for proton or sodium groups on the resin, or a base to elute anions that have been exchanged for hydroxide or chloride groups on the resin, to reverse the equilibria as presented in Equations 21.1-21.3. See Chromatography on Classical Ion Exchange Resins, Section 21.2.

5. Gel type resins are solid polymer beads (fairly large) with no appreciable porosity. They are typically used for water purification rather than analytical separations. Macroreticular resins are porous, having greater surface area. . See Chromatography on Classical Ion Exchange Resins, Section 21.2.

6. Their functional groups differ as follows. Strong acid: sulfonic acid; weak acid: carboxylic acid; strong base: quarternary ammonium ion; weak base: amine group. See Table 21.1 for commercial types.

7. Cation impurities are exchanged for protons and anions for hydroxide ions, resulting in H_2O in place of the salts. So the specific conductance will be that of water, 54.84 nS/cm (see Example 21.1).

8. See Figure 21.7 and the discussion thereof illustrating several types of modern ion exchange phases. Many based on chemically modified latex particles of different porosity with various coatings.

9. A monolithic column consists of a single solid rod, permeated throughout with interconnecting pores, instead of packed particles. They have lower pressure drops and allow faster flow rates. They exhibit efficiency comparable to 3.5 μm packed particle columns. They were first described by Stellan Hjertén of the University of Upppsala, who also pioneered the use of agarose gels, made seminal contributions to monolithic columns, and first demonstrated free solution zone electrophoresis in small quartz tubes. See Section 21.2, A monolithic column, and Section 21.10.

10. The zwitterionic phase may be bonded at the + or − end. For analytes with significant electrostatic interaction, thee selectivity will depend on this orientation. See Stationary Phases for Hydrophilic Interaction Chromatography, Section 21.2.

11. Chiral stationary phases include Pirkle, chiral cavity, helical polymer, and ligand exchange types. See the descriptions of each in Section 21.2 for the different mechanisms. See Chiral Stationary Phases, Section 21.2.

12. Alumina, zirconia, titania, and porous graphitic carbon withstand wide pH range, with crosslinked bonded phases. The substrates and phases are also thermally stable. See Improving hydrolytic stability at pH and temperature extremes, and Other Supports, Section 21.2.

13. A solvent delivery system requires an inlet filter, a solvent degassing system, and a pulse dampener, in addition to the pump. See Solvent Delivery System, Section 21.3.

14. The main pump systems are positive displacement pumps, the most common the reciprocating pump, and syringe pumps. See Pumps for Solvent Delivery, Section 21.3.

15. A low pressure gradient pump is used when the solvent composition changes on the feed side, while a high pressure one is used when on the output side. A quarternary pump is used to run a gradient containing up to four components. See Pumps for Solvent Delivery, Section 21.3.

16. In the reciprocating pump, the piston and the seat of the check valve are made of sapphire, and the ball of the check valve of ruby. See Pumps for Solvent Delivery, Section 21.3.

17. The injector consists of a stator the which external connections are made, and an internal disc-shaped rotor to switch positions of ports relative to one another, and therefore the flow of solutions. It has four external ports. See Figure 21.5 for operation steps. See Sample Injection System, Section 21.3.

18. In RPC, the stationary phase is nonpolar and the eluting solvent polar. Nonpolar compounds are most strongly retained and less soluble in the solvent. The elution order will be benzoic acid, acetone, benzene. See HPLC Subclasses, Section 21.1.

19. The most common RPC bonded phase is octyldecyl silica (ODS), known as C18. Others include phenyl, biphenyl, diphenyl, and C8 (less hydrophobic than C18). HILIC phases are neutral (amide or diol bonded to porous silica), charged (bare silica or amino, aminoalkyl or sulfonate bonded to porous silica), and zwitterionic. See HPLC Subclasses, Section 21.1, Section 21.2 and Stationary Phases for Hydrophilic Interaction Chromatography.

20. Unreacted silanol groups (-SiOH), which may provide polar interaction sites, are endcapped with, e.g., $-CH_3$, to minimize this. See Section 21.2.

21. See Figure 21.5 for structures. Microporous particles have small pores, and the solute must diffuse in and out of these. Perfusion particles have a mixture of large and small pores, allowing the solution to flow directly through the particles, increasing mass transfer. They can be used at higher flow rates and have better efficiencies for biomolecules. Nonporous particles have a solid core, with a thin porous layer. They are small and eliminate stagnant mobile phase, giving increased mass transfer, but at the expense of higher backpressure and limited capacity. They are used in ion chromatography.

22. The guard column retains debris and sample particulate matter, and it retains highly sorbed compounds that would be caught on and not be eluted from the analytical column. It extends the life of the analytical column. Guard columns must be replaced periodically, and generally are not used when highly efficient columns packed with small particles where extra-column dispersion becomes all-important. See Column, Section 21.3.

23. UHPLC uses smaller particles (STM, sub-two micron), and narrower, and shorter columns, operated at higher velocities to achieve faster separations, with reduced solvent consumption. With small particles, the van Deemter H vs. u plot becomes flat at high flow rates because small particles are less resistant to mass transfer. See Section 21.6.

24. In gradient elution, the polarity of the solvent mixture is varied with time, increasing the fraction of the strong solvent. This allows weakly retained compounds to be eluted later and better resolved, and longer retained ones to be eluted more quickly. The comparable thing is achieved in gas chromatography by temperature programming. See Section 21.5.

25. Narrowbore columns give narrower, taller peaks and increased sensitivity. Or smaller sample can be injected with the same sensitivity.

26. The stationary phase is a molecular sieve that has an open network, which excludes solutes above a certain size. Solutes of formula weight above the limit are not retained by the column and can be separated from smaller solutes that are retained. The exclusion limit is the formula weight of the molecule that is excluded (not retained) by the molecular sieve. See Section 21.1, HPLC Subclasses.

27. Size exclusion chromatography separates molecules based on size. Those larger than the pore size of the stationary phase are excluded and are eluted together. Those smaller than the smallest pore size come out last. Intermediate ones are in between. It is used in biochemistry for low pressure/gravity-fed preparative separation of biomolecules. Ion exclusion chromatography is used to separate organic acids from strong acids, based on a cation exchange resin that excludes the anion, but not the unionized acid. See Section 21.1, HPLC Subclasses.

28. In Size exclusion chromatography, molecules are separated based on their size. The stationary phase volume is largely occupied by pores, and molecules larger that these are excluded. In micellar electrokinetic chromatography, micelles act as a pseudo-stationary phase in a capillary electrophoresis experiment. Neutral solutes will partition between the micelles and buffer, and different molecules interact differently. The electrophoretic migration of the negatively charged micelles is toward the anode, but the net flow is toward the cathode, due to electroosmotic flow, but at a slowed rate. Neutral molecules move at the electroosmotic flow rate, but when they interact with the micelle, their movement is slowed. See Section 21.1, HPLC Subclasses, and Section 21.11.

29. Universal detectors include the mass spectrometer, and refractive index and viscosity detectors. The RI changes with temperature and is incompatible with gradient elution. Viscosity detectors have low sensitivity. See Section 21.3, Detectors.

30. The RI detector is particularly useful for solutes that have little or no useful UV absorption, e.g., carbohydrates. Viscosity detectors are useful for polymer detection. Light scattering detectors can provide molecular weight distribution and size information. Most RI detectors are based on a differential refractometer design, and measure the difference in the index of refraction of the solvent and the solvent plus solute. See Section 21.3, Detectors.

31. All dissolved substances are either charged or uncharged, and so the electrical conductivity detector will detect charged ones, such as charged colloids. Aerosol detectors are useful for all solutes that are non-volatile and hence form aerosol particles when the column effluent is nebulized in a dry gas. See Section 21.3, Detectors.

32. Friedrich W. G. Kohlrausch (Margin picture Isotachophoresis). Application of a DC potential in conductivity measurements can lead to undesirable electrochemical reactions at the electrodes. But with high background resistance, as in ion chromatography DC measurements provide simple inexpensive detection. See Section 21.3, Detectors.

33. HPLC detectors should have low noise and high sensitivity for good detection limits. See Section 21.3, Detectors.

34. From the Hagen-Poiseuille equation (21.10), the pressure drop is inversely proportional to r^4. Decreasing r by a factor of 2, then, increases the pressure drop 16-fold.

35. In the capacitive coupled contactless conductivity detector, a pair of ring-shaped electrodes are placed on the outside of a capillary and a high frequency voltage applied, sufficient to

penetrate the capillary walls. The conductance of the solution is measured, using a current-to-voltage converter connected between one electrode and ground. The rectified signal is proportional to the solution conductance. The advantage for microscale systems is avoiding making measurements in a separate cell, or having to insert a wire into the capillary. See Section 21.3, Detectors.

36. In aerosol detectors, the effluent is passed into a nebulizer, and minute particles form from the nonvolatile analytes, which pass through a focused light beam, and the light scattered by the particles is registered. See Section 21.3, Detectors.

37. In suppressed ion chromatography, the eluted analyte and the eluent emerging from an analytical ion exchange column are passed through a high capacity suppressor column (cation exchanger, H^+ form for base eluent; anion exchanger, OH^- form for acid eluent) to convert the eluent to H_2O or a low conducting acid. This allows sensitive continuous conductometric detection of the analytes as they emerge separately, to provide an ion chromatogram. Nonsuppressed IC uses a single column. If there is a sufficient difference in the equivalent conductance of the eluent ion (e.g., OH^-) and the analyte ion which displaces it (e.g., SO_4^{2-}), a negative shift in conductance is registered. See Section 21.4.

38. Like nonsuppressed conductivity detection (see 37 above), indirect photometric detection in HPLC relies on difference in absorbance of the background solution and the analyte. See Section 21.10, Detectors in CE.

39. Packed column suppressors can tolerate high pressures, but require regeneration and switching. Ion exchange membrane-based suppressors can be operated continuously, without intermittent operation. See Section 21.4, Membrane Devices.

40. In chemically regenerated columns, the NaOH regenerant always contains impurities such as NaCl and NaClO$_3$, and CO_2 in the atmosphere makes it hard to avoid forming some carbonate. Electrodialytic membranes avoid these problems by generating the hydroxide eluent. See Section 21.4, Electrodialytic Membrane Devices.

e 21.22). Carbonate is removed by a continuously regenerated ion trap column (Figure 21.23).

42. KOH is used because K+ has the highest mobility, leading to the lowest voltage drop and the least Joule heating, and allows the largest eluent concentration to be suppressed. See Section 21.4, Membrane Devices.

43. The LCW has less light loss per unit length that in a standard cylindrical tube. Liquid core waveguides are good for flow-through fluorescence detection. Light is incident transversely on a LCW tube through which the analyte bearing solution is flowing. Any unabsorbed light passes out in the radial direction, while fluorescence is in all directions. See Figure 16.34 for details.

44. Variable wavelength UV-visible detectors, e.g., photodiode array detectors, are the most commonly used in HPLC instruments. See Section 21.3, Detectors. See Chapter 16, Figure 16.25: diagram of a photodiode array detector. The only difference is that the optical cell operates in a flow-through configuration, such as the flow cell shown in Figure 16.16.

45. Ratio absorbance plotting involves measuring the absorbance at two different wavelengths, and plotting the ratio of them. An analyte will have a unique ratio. If the ratio at an analyte peak is not as expected, this suggests coelution. See Section 21.3, Detectors.

46. See Section 16.10, Figure 16.25 for an array detector. In HPLC, it is equipped with an appropriate flow cell (Section 21.3, Detectors).

47. Electrochemical detectors are useful for analytes that can be reduced or oxidized. The coulometric detector is useful for nitro polynuclear aromatics, with nitro group being reduced. The pulsed amperometric detector is useful for detecting carbohydrates containing oxidizable –OH groups. See Section 21.3, Detectors.

48. In postcolumn detection, a chemical reaction occurs after the analyte elutes from the column to give a detectable species prior to the detector. Analyte conversion can also be accomplished by UV irradiation, heating, or by passing through a solid reactor. The two important parameters in PCR detection are mixing efficiency and time for reaction. See Section 21.3, Postcolumn Reaction.

49. Post column reactor designs include tubular reactors, the arrow mixer, knitted tubes, and pearl string reactors. Gas –segmentation is used for long reaction times to minimize dispersion. See Section 21.3, Postcolumn Reaction.

50. A typical Ion Chromatograph generates the eluent electrodialytically (typically a hydroxide eluent), purifies it (removes carbonate from CO_2 that may have been present in the water) using an electrodialytically generated trap column and removes the eluent counterion by using an electrodialytic suppressor. No other LC technique does anything comparable.

51. An eluent generator produces electrolytic gas. When a hydroxide eluent is generated, H_2 gas is formed in the eluent which is removed by a gas permeable membrane [p 79-80]. The suppressed product of a carbonate eluent is H2CO3; most of the CO2 present can be removed and the background thus reduced by removing the CO2 through a gas permeable membrane before the detector.

52. In gradient elution, the column has to be re-equilibrated, taking time. An isocratic elution may provide faster measurements. See Section 21.5.

53. Small bore columns can be used at high linear velocities to achieve a fast separation. And less solvent is consumed. See Section 21.4.

54. In any type of exclusion chromatography, e.g., size exclusion chromatography, ion exclusion chromatography, etc.

55. As with the retention factor in HPLC, the R_f value in TLC is characteristic for a given stationary phase-solvent composition. The largest value is unity, the smallest value zero See Section 21.6.

56. In HPTLC, smaller particles are used, with thinner sorbent layer thicknesses. The migration distance is shorter, and separation times are shorter. Band broadening is reduced. HPTLC is useful when a large numbers of analyses are to be conducted. Stationary phases are typically unmodified silica. See Section 21.6. Stationary Phases.

57. For best reproducibility, automated instrumental sample application using the spray-on technique is preferred. Common ways of for exposing the plate to a liquid reagent are spraying the reagent on the plate or immersion in the reagent. An equivalent of indirect photometric detection is to bind a fluorescent marker to the stationary phase. Under UV illumination, the nonfluorescent analytes are visualized as dark spots. See Section 21.6, Sample Application, Visualization.

58. A typical chamber may not be saturated with the developing solvent vapor, and the composition of the gas phase can differ significantly form that of the developing solvent, depending on the relative vapor pressures of the different solvent components. The mobile phase composition changes as it moves up the plate. The R_f values are larger than in a saturated, preconditioned chamber. See Section 21.6, Developing the Chromatogram.

59. In HPTLC, stepwise gradient elution is done using Automated Multiple Development. The plate is developed repeatedly in the same direction and stopped when the solvent front moves a certain direction. The plate is dried and a next development is is done with a less polar solvent. . See Section 21.6, Developing the Chromatogram.

60. Visualization is achieved by different methods, from the color of the spot, by using a fluorescent background to create a dark spot of the analyte, etc. Quantitative measurements include photographic imaging in the visible region, variable wavelength reflectance scanners, and MALDI-MS measurement. See Section 21.6, Detection of the Spots and Quantitative Measurements.

61. The sample travels through a capillary by means of electroosmosis, by inserting the ends of the capillary in a buffer solution and applying a high D.C. voltage by means of platinum electrodes immersed in the solution. Uncharged analytes migrate together at the electroosmotic rate, but charged analytes travel at different rates, based on their electrophoretic mobilities. See Section 21.10 and Figure 21.28.

62. DNA fragments in particular are built of similar repeated sequences. As the size of the fragment increases, the charge increases therefore proportionally and the charge/size ratio that governs mobility remains the same. Movement through a gel, however, is not unrestricted and the length of the chain matters.

63. In slab gel electrophoresis, a voltage is applied in one direction. In pulsed field gel electrophoresis, the voltage is applied in one direction for several seconds, then reversed for a shorter time, repeating. It finds use in measuring DNA fragments, which may migrate in a non-predictable manner unrelated to size. In this technique, large DNA fragments are slowed down relative to the smaller ones, which can change direction faster. See Section 21.9, Related Techniques.

64. In planar gel electrophoresis, field strengths of 4-6 V/cm are common. In CE, several thousand volts are applied, e.g., 500V/cm. The plate numbers and resolution in CE are proportional to the applied voltage, see Equation 21.28 and Example 21.8. See Section 21.9 and Section 21.10.

65. Isoelectric focusing is used to separate proteins which have no net charge at the isoelectric point, but charge varies with pH. In IEF, the separation medium is an immobilized pH gradient gel. As the protein travels through the gel of increasing pH, it reaches its isolectric point pH and stops moving, and the proteins focus into individual well-defined bands. Resolution in capillary isoelectric focusing (CIEF) can distinguish proteins differing by a single protolyzable group. See Section 21.9.

66. PAGE refers to polyacrylamide gel electrophoresis. Proteins are often treated with the anionic detergent sodium dodecyl sulfate (SDS), which linearizes the protein molecule and imparts a negative charge to it, and proteins are separated by size using SDS-PAGE. See Section 21.9.

67. In pressure-driven HPLC, laminar flow occurs, in which the fluid velocity is zero at the wall and twice the mean velocity at the center, resulting in a Gaussian band, and lower amplitude. In CE, the analyte moves electrophoretically and via the EOF. The radial electric field profile is flat, and the ions electrophoretically move like a plug, resulting in high resolution. However, CE is not so robust as HPLC. Any substance that adsorbs on the wall will change the EOF, a major problem with real-world samples. See Section 21.10, How CE Can Provide High Efficient Separations, and Electrophoretic Mobility and Separation.

68. Ethidium bromide is highly mutagenic. It is also a UV-excited fluorophore and unless proper protection is taken, UV radiation can cause eye damage.

69. ZIFE is zero integrated field electrophoresis, for separating large DNA fragments. The voltage forward and reverse time steps are equal. Separation occurs because of the asymmetric rate at which the molecules reverse their movement. See Question 63 for more information. See Section 21.9, Related Techniques.

70. Commercial CE instruments provide an interlock. Whenever the instrument cover is opened, the HV source is disconnected. See Figure 21.29.

71. The two main modes of injection in CE are pressure-based injection and electrokinetic injection. In the former, all ions are injected without preference for any particular ion. In the latter, ions are differentially injected due to the applied electric field. Thus, with a positive

injection potential, cations will migrate into the capillary in preference to anions. This may complicate quantification, but can be advantageous as ions can be concentrated based on their mobililties. See Section 21.10, Operation.

72. In electrostacking, the background electrolyte has a higher ionic strength than the sample solution. The migration velocity depends on the ion mobility. When a voltage is applied, the sample zone exhibits higher resistance than the electrolyte, producing a higher field strength, and causing the analyte ions to migrate faster than the electrolyte ions. This differential flow rate results in accumulation of a very narrow sample zone at the sample-carrier electrolyte boundary. This is useful for large volume injections of low conductivity samples, enhancing sensitivity. See Section 21.10, Operation.

73. James Jorgenson at the University of North Carolina first observed EOF, and also invented UHPLC. EOF originates from the fact that fused silica capillaries have ionizable silanol groups, -SiOH, which at all but very acid pH are ionized to –SiO⁻, creating a negatively charged (zeta potential) surface. This attracts cations from the buffer solution to create an electrical double layer along the walls. A diffuse layer of the cations extends toward the center of the capillary. When a high DC voltage is applied, the mobile-phase positive charges in the diffuse layer migrate in the direction of the cathode. Because the ions are solvated, the buffer fluid is dragged along by the migrating charge, creating a solution flow of the bulk solvent. See Section 21.10, How CE Can Provide Highly Efficient Separations, and Figure 21.31.

74. Actually, the EOF may be too fast or too slow or too pH dependent for an optimum separation, and in many cases it may be best to have no EOF all. The capillary may be coated with a sulfonated polymer for a pH-independent EOF, or with an immobilized neutral material to obtain a near zero zeta potential and hence a zero EOF. See Section 21.10, How CE Can Provide Highly Efficient Separations.

75. Because the capillary dissipates the heat on the outer surface, the center of the fluid is the hottest and there is a radial temperature gradient that is proportional to the total power dissipated. It is also proportional to the square of the capillary inner diameter and inversely as the thermal conductivity of the fluid. Temperature affects viscosity and changes both electrophoretic mobility and EOF. The electrophoretic mobilities of common ions typically change about 1.7%/°C. If the ions or the bulk fluid is moving faster at the center of the capillary than at the walls, we approach the same situation that we have in laminar flow. In addition, the fluid obviously leaves hotter than it enters. So there must be an axial temperature gradient as well.

76. The zeta potential is that created on the capillary wall from the ionized silanol groups, being pH dependent. See Section 21.10, How CE Can Provide Highly Efficient Separations.

77. Reversed phase chromatography, with a polar stationary phase, would be closest to MEKC with micelle pseudo-stationary phase. MEKC is useful for separating water-insoluble neutral compounds, such as biological substances like sterols. See Section 21.11.

78. In isotachphoresis, ions stack according to their mobilities., the most mobile towards the leading electrolyte. The isotachophoretic condition is the characteristic that once steady state is reached, all bands move at the same velocity, given by Equation 21.35. See Section 21.11.

79. The partition constant (see partition coefficient, Equation 19.2) increases as the solubility increases in the stationary phase. Since a C18 (nonpolar) column is used the compounds that elute first have the lowest solubility. So the partition constant order is GSSG>GSH>Hcyss.

80. (a) By using the equation 19.8

$$N = 16(\frac{t_R}{W_b})^2$$

The effective number of theoretical plates for the column is calculated to be:

$$N_A = 16(\frac{15.80 - 1.60}{1.25})^2 = 2065$$

$$N_B = 16(\frac{17.23 - 1.60}{1.38})^2 = 2052$$

So the average N=(2065+2052)/2=2058

(b) From equation 19.5, the plate height H=L/N=30.0 cm/2058=0.014 cm

(c) By using equation 19.31

$$R_s = \frac{2(t_{R_B} - t_{R_A})}{W_{b_B} + W_{b_A}}$$

The resolution of A and B is thus calculated to be:

$$R_s = \frac{2(17.23 - 15.80)}{1.25 + 1.38} = 1.09$$

(a) Because R_s^2 is proportional to L or N (equation 19.34), to get a resolution of 1.5, the new column length must be $(1.5/1.09)^2 = 1.89$ times longer. The almost double required column length is thus 30 cm x 1.89 = 57 cm

(b) Under identical elution conditions, t_R values are linearly related to column length, thus

$$t_{R_A} = 15.80 \text{ min x } 1.89 = 29.86 \text{ min}$$
$$t_{R_B} = 17.23 \text{ min x } 1.89 = 32.56 \text{ min}$$

81. mmol K^+ = mmol H^+ = mmol NaOH
    mmol K^+ = 0.0506 mmol/mL x 26.7 mL = 1.35 mmol
    1.35 mmol/5.00 mL = 270 mmol/L

82. meq = mmol for a monovalent ion
    (10 g/L)/(58 g/mol) = 0.17 mol/L = 170 mmol/L
    170 mol/L x 0.20 L = 34 mmol Na^+ = 34 meq
    (34 meq)/(5.1 meq/g) = 6.7 g resin

83. (a) HCl (b) H_2SO_4 (c) $HClO_4$ (d) H_2SO_4

84. (a) The flow cells need to be transparent to the incident radiation. IR "sees" vibrations of molecular bonds so the cells have to be made of highly transparent materials with no molecular bonds (e.g., salts).

    (b) The vast majority of HPLC applications are performed by reversed phase chromatography, which uses aqueous mixtures. IR cell material (e.g., NaCl) are not impervious to these kinds of mobile phases.

    (c) The mobile phase should be transparent to the incident radiation because there is so much more of it than analyte. It is hard to find a region in the IR spectrum that an analyte might absorb that most potential mobile phase components do not.

    (d) Typically, IR bands (e.g., C-H stretch region) are comprised of overlapping bands, all with slightly different intensities. Hence, IR does not always obey Beer's Law.

    (e) IR spectrometry is generally not as sensitive as UV.

85. Capillaries have small lumens. Even if an external cell with a Z configuration is used to increase optical path length it is difficult to get long enough path for good detection limits and wide enough diameter to allow much light to reach the detector, another source of S/N degradation. Fluorescence detection is usually performed with a tightly focused light source, which may even be a low power laser. Fluorescence is generated from a small volume and the diameter of the capillary is usually irrelevant.

86. If the peak half width is 3s, the total width is about 6 s. If 20 measurements must be made in 6s, the minimum data rate should be 20/6 = 3.3 Hz.

    5 Hz would be the best option.

87. According to equation 21.21

$$[\alpha]^T_\lambda = \frac{\alpha^T_\lambda}{cl}$$

where $[\alpha]^T_\lambda$ is the specific rotation at temperature T and wavelength λ; l is the optical path length in dm; α = observed optical rotation, c = concentration in g/mL.
According to the text, good polarimetric detectors have noise levels of 20 □deg. The limit of detection is typically taken to be three times the noise above the background; in this case therefore this will be a signal of 6 x 10^{-5}. Here l = 0.2 dm.

Therefore, at the detection limit, putting $\alpha_\lambda^T = 6 \times 10^{-5}$, we have:

$$c = \frac{\alpha_\lambda^T}{[\alpha]_\lambda^T l} = 6 \times 10^{-5}/(49 \times 0.2) \text{ g/mL} = 6.1 \times 10^{-6} \text{ g/mL or } 6.1 \text{ µg/mL}$$

Accounting for a chromatographic dilution factor of 10, the concentration LOD is 10-fold higher, 61 µg/mL.

88. Energy loss $E = \dfrac{hc}{\lambda} = hc \,(3382 \text{ cm}^{-1}) = 3.382 \times 10^{-4} \, hc \text{ nm}^{-1}$

Excitation photon energy $= \dfrac{hc}{253.7} \text{ nm}^{-1} = 3.942 \times 10^{-3} \, hc \text{ nm}^{-1}$

Emission photon energy $= (3.942 - 0.338) \times 10^{-3} \, hc \text{ nm}^{-1} = 3.604 \times 10^{-3} \, hc \text{ nm}^{-1} = \dfrac{hc}{277.5} \text{ nm}^{-1}$

The emitted wavelength is 277.5 nm

89. In liquid chromatography, the area of an optical absorption peak is approximately the product of the peak height (in absorbance units, directly proportional to the product of the molar concentration of the analyte times the molar absorptivity) and the half peak width in volume units. Normally the abscissa (x-axis) of the chromatogram is given in time units – but it is easily converted to volume by multiplying with the flow rate. The peak area is thus directly proportional to the moles of the analyte. For a negative peak that results from an absorbing eluent ion being replaced by a <u>nonabsorbing</u> analyte ion, the effective (negative) absorptivity is the same for all analytes and is equal in magnitude to the eluent absorptivity. For a discussion, see S. A. Wilson and E. S. Yeung, Anal. Chim. Acta 157(1984) 53–630. This paradigm works in liquid chromatography because the detector is outside the column and the flow rate (not normally varied during a chromatogram) of any analyte through the detector is the same.

A slightly more involved consideration is needed for CE because the detector is typically within the electric field and analytes move through the detector at different velocities. And therefore the absorbance*time product has to be multiplied by μ_{net} or u_{net} ($u_{net} = \mu_{net} * E$) which is simply linearly related to the reciprocal of the migration time t_m. We still need, however, one yardstick. If any solute A of known concentration is injected and the calibration constant K is computed from $C_A = K$ (peak area in AU.s)$/t_{m,A}$ then this can be used to compute the concentration of all other analytes.

90. From Conversion tables: 1 psi = 6.89×10^5 dynes/cm^2

Based on the Hagen-Poiseuille equation 21.10,

$$F = \frac{DP\pi r^4}{8\eta L}$$

where η, the viscosity of water, is 8.90×10^{-3} dyne s/cm^2, equaling to 1.29×10^{-7} $\Delta P = 5\,psi$ $= 5 \times 6.89 \times 10^5$ dyne/cm$^2 = 3.44 \times 10^6$ dyne/cm^2 at 25 °C. The inner diameter of the capillary is 75 μm or 7.5×10^{-3} cm. The radius r is therefore $3.7_5 \times 10^{-3}$ cm.

$$F = \frac{3.44 \times 10^6 \text{ dyne cm}^{-2} \times 3.14 \times (3.75 \times 10^{-3})^4 \text{cm}^4}{8 \times 8.90 \times 10^{-3} \text{ dyne s cm}^{-2} \times 50 \text{ cm}} = 0.60 \times 10^{-3} \text{ cm}^3/s = 6.0 \times 10^{-7} \text{ L/s} = 60 \text{ nL/s}$$

In 2 sec injection, 120 nL will be injected.

The sample amount ratio for electrokinetic introduction is

$$r = \lambda_i C_i / \lambda_j C_j$$

If equal amounts are introduced, r is 1.

$$\lambda_i C_i = \lambda_j C_j$$
$$\frac{C_i}{C_j} = \frac{\lambda_j}{\lambda_i}$$

Using the equivalent conductance data in Table 21.2, chloride and iodate are 76.35 and 40.5 S cm^2 equivalent^{-1}.

$$\frac{C_{Cl}}{C_{IOs}} = \frac{\lambda_{IOs}}{\lambda_{Cl}} = \frac{40.5}{76.35} = 0.53$$

91. It is first necessary to ascertain the μ_{net} of each ion. The μ_{eo} is given as 9×10^{-4} cm^2 V^{-1} s^{-1} and the μ_{ep} determined for chloride and iodate from equation 21.13 and table 21.2 are -7.91×10^{-4} and -4.20×10^{-4} cm^2 V^{-1} s^{-1}, respectively. The $\mu_{net} = \mu_{eo} + \mu_{ep}$, and μ_{net} values for Cl$^-$ and IO$_3^-$ are 1.08×10^{-4} and 4.80×10^{-4} cm^2 V^{-1} s^{-1}. Thus, iodate has a net velocity 4.80/1.08 = 4.44 times higher than chloride.

If the peak areas in terms of absorbance units * time are equal, the iodate peak area will 4.44 times that of chloride if we calculate the peak areas in terms of absorbance units* volume. Iodate is thus present in 4 times the concentration as chloride; in CE, the apparent peak areas can be misleading.

92. See Section 21.3, Sample Injection System for description of loop type injection valves.

(a) This is an exact equivalent of a 6-port injector: It can also be done with 5:

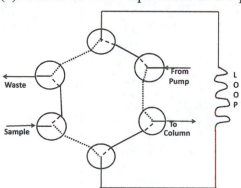
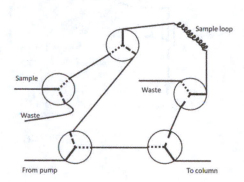

With a tee, it is possible to do with 3:

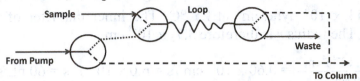

Or with 2, if there are two tees and pumped liquid can flow through a restricted by pass when sample is being loaded in loop:

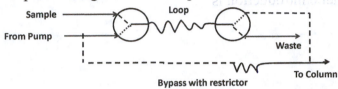

Or with no tees and no bypass if pump can be stopped when sample is loaded:

93. See the Valco Instruments Inc (VICI) website for a description:
http://www.vici.com/support/app/app14j.php

Simultaneous Injection of the Same Sample onto Separate Columns

Click below to change the valve position. (Currently shown in Position A)
Position A
Position B

A: Loops 1 and 2 filled in series

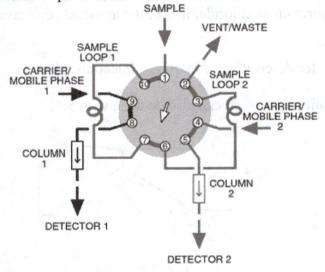

In Position A, sample fills the two loops in series. In Position B, the sample is simultaneously injected into two separate flow systems. A single autosampler used with this flowpath can automate two analytical procedures for the same sample.

In an important non-chromatographic application, the roles of carrier and sample are reversed, permitting two different quantities of two different materials to be dispensed together, as in automatic dilution.

94. Based on the Hagen-Poiseuille equation,

$$\Delta P = \frac{8F\eta L}{\pi r^4}$$

The viscosity η of 50% methanol solution at 30 °C is 1.43 centipoise or 1.43×10^{-2} dyne s/cm^2

$F = 100 \ \mu L/min = 0.1/60 \ cm^3/s = 1.67 \times 10^{-3} \ cm^3/s$
$L = 1 \ m = 100 \ cm$
$r = 50 \ \mu m = 5 \times 10^{-3} \ cm$

$$\Delta P = \frac{8 \ ' \ 1.67 \ ' \ 10^{-3} cm^3 \ s^{-1} \ ' \ 1.43 \ ' \ 10^{-2} \ dyne \ s \ cm^{-2} \ ' \ 100 \ cm}{3.14 \ ' \ (5 \ ' \ 10^{-3})^4 \ cm^4} = 9.73 \times 10^6$$

dyne/cm^2 = 0.973 MPa = 141 psi

The viscosity of 50% acetonitrile solution at 30 °C is 0.74 centipoise, about 0.517 times that of the methanol. The pressure is thus going to be 0.973 x 0.521 MPa = 0.507 MPa = 73.5 psi

95. Consider 0.02 □M LiOH. We have Li$^+$, H$^+$, and OH$^-$ in equilibrium, each contributing to the conductivity. From Table 21.2, the limiting ionic conductances (S cm^2 equiv^{-1}) are: Li$^+$ 38.69, H$^+$, 349.92, OH$^-$ 198.6.

For 2.0×10^{-8} equiv/cm^3 LiOH, the pH is determined by H$^+$ water ionization in the presence of the dilute LiOH, and we must use the quadratic equation to calculate, giving H$^+$ = 9.05 x 10^{-8} equiv/cm^3 (pH is 7.107), and OH$^-$ = 1.1 x 10^{-7} equiv/cm^3. Hence, from Equation 21.16,

σ_c = [Li$^+$] equiv/cm^3 x 38.69 S cm^2 equiv^{-1} + [H$^+$] equiv/cm^3 x 349.92 S cm^2 equiv^{-1} +[OH$^-$] equiv/cm^3 x 198.6 S cm^2 equiv^{-1}

σ_c = (2.0 x 10^{-8} x 38.69 + 9.05 x 10^{-8} x 349.92 + 1.1 x 10^{-7} x 198.6) x1000 = 0.0544 μS/cm, or 54.4 nS/cm.

A similar calculation for pure water gives 54.8 nS/cm. The contribution from the highly mobile H$^+$ in the presence of the LiOH is decreased, and we see that the conductivity can be less than that of pure water. However, the difference between them is less than 1%.

LiOH(μM)	0	0.005	0.010	**0.020**	0.050	0.100
σ_c (nS/cm)	54.8	54.7	54.5	**54.4**	54.7	57.6

96. According to equation 21.16,

$\sigma_c = \Sigma \lambda_i C_i$

where σ_c is the expected conductance and λ_i is the equivalent conductance of each ion and C_i is the concentration of the ion in eq/L.

Therefore,

$\sigma_c = (10^{-3}$ equiv/L x 73.5 S cm^2/equiv + 10^{-3} equiv/L x 76.35 S cm^2/equiv)

Putting 1 L = 1000 cm^3,

$\sigma_c = 149.85$ x 10^{-6} S cm^2/cm$^3 = 149.85$ μS/cm

97. (a) According to Table 21.2, λ_{K+} and λ_{Cl-} are respectively 73.5 and 76.35 S cm^2 equivalent^{-1}.

Based on equation 21.17, at 298 K

$D = \dfrac{\lambda RT}{zF^2}$

For K$^+$

$D_{K+} = (73.5$ S cm^2 equiv^{-1} * 8.31 J K^{-1} equiv^{-1} * 298 K) /(1 * (96485 coulomb equiv$^{-1})^2)$
$\quad = 1.95$ x 10^{-5} S cm^2 J C^{-2}

Recognize Siemens S = amp/volt
Joule J = Watt. s = Volt*amp .s
Coulomb C = amp.s

Thus

$D_{K+} = 1.95$ x $10^{-5} \dfrac{amp}{volt}$ x cm^2 x $Volt * amp * s$ x $\dfrac{1}{(amp * s)^2} = 1.95$ x 10^{-5} cm^2/s

Thus

$D_{Cl-} = 1.95$ x 10^{-5} cm^2/s * $(\lambda_{Cl-}/\lambda_{K+}) = 1.95$ x 10-5 *76.35/73.5 cm^2/s = 2.03 cm^2/s

(b) The Stokes Radius could be calculated using the equation

$$r = \dfrac{kT}{6\pi\eta D} \quad (21.11)$$

The viscosity of water (η) at 298K of 8.94 x 10^{-3} g cm^{-1} s^{-1}

Therefore,

$$r_{K+} = \dfrac{1.38 \times 10^{-23} \text{ J K}^{-1} \times 298 \text{ K}}{6\pi \times 8.94 \times 10^{-3} \text{ g cm}^{-1} \text{ s}^{-1} \times 1.95 \times 10^{-5} \text{ cm}^2 \text{ s}^{-1}}$$

$$= 1.25 \times 10^{-15} \text{ Js}^2 \text{ g}^{-1} \text{ cm}^{-1} = 1.25 \times 10^{-15} \dfrac{\text{J cm}}{\text{erg}} \times \dfrac{10^7}{\text{J}} \text{ erg} = 0.125 \text{ nm}$$

$r_{Cl-} = r_{K+}(D_{K+}/D_{Cl-}) = 0.125*1.95/2.03$ nm $=0.120$ nm

98. As the wires are small relative to the capillary diameter, as a first approximation we can consider the cell as two 10 μm dia. disks separated by 50 μm. The area A of the 10 μm diameter electrode face is
$A = \pi(10^{-3})^2/4 = 7.85 \times 10^{-7} \text{ cm}^2$
If L = 0.005 cm, then $\kappa = L/A = 0.005 \text{ cm} / 7.85 \times 10^{-7} \text{ cm}^2 = 6.37 \times 10^3 \text{ cm}^{-1}$

99. From example21.1, pure water only contains H^+ and OH^-, each at a concentration of 10^{-7} eq/L, that we can express as 10^{-10} eq/cm^3.

   Thus (Equation 21.16),
   $\sigma_c = \lambda_{H+}[H^+] + \lambda_{OH-}[OH^-] = 349.82 \text{ S cm}^2 \text{ eq}^{-1} \times 10^{-10} \text{ eq cm}^{-3} + 198.6 \times \text{S cm}^2 \text{ eq}^{-1} \times 10^{-10} \text{ eq cm}^{-3} = 54.84 \text{ nS/cm}$.

   For pure water, From equation 21.14,
   $G = 54.84 \text{ nS/cm} / 6.37 \times 10^3 \text{ cm}^{-1} = 8.61 \text{ fS}$

   1 mM KCl contains $[K^+]$ and $[Cl^-]$ in concentrations of 10^{-3} eq/L, or 10^{-6} eq/cm^3. Thus $\sigma_c = \lambda_{K+}[K^+] + \lambda_{Cl-}[Cl^-] = 73.5 \text{ S cm}^2 \text{ eq}^{-1} \times 10^{-6} \text{ eq cm}^{-3} + 76.35 \times \text{S cm}^2 \text{ eq}^{-1} \times 10^{-6} \text{ eq cm}^{-3} = 149.85 \text{ μS/cm}$.

   So for 1 mM KCl, $G = 149.85 \text{ μS/cm} / 6.37 \times 10^3 \text{ cm}^{-1} = 23.5 \text{ nS}$

100. KCl is available in highly pure form, is relatively inexpensive and consists of relatively high mobility ions, thus forming a highly conductive solution at high concentrations.
   Electrolytes like KCl are often used in a concentrated form inside a reference electrode, which makes a contact with the outer solution via frit. Here it has the great advantage over its more common cousin NaCl in that the cation and anion has almost the same mobility or diffusion coefficient, and a large junction potential is not developed because one ion diffuses away faster. In the case of NaCl, Cl^- diffuses through the frit at a faster rate than Na^+, resulting in the inner more concentrated solution side of the frit developing a positive potential relative to the outside.

101. Arginine has the shortest migration time because it has the most positively-charged side chain at pH 10; thus, its net migration toward the cathode is faster than that of the other amino acids.

102. (a) From Equation 21.13, the electrophoretic mobility is inversely proportional to the charge. Hence, the molecule with one charge has the highest value. (b) It will have a negative value since Z is negative. See also the calculations in Examples 21.6 and 21.7.

103. $\mu_{net} = \mu_{ep} + \mu_{eo}$
   At pH 9.00, $\mu_{eo} = 3.1 \times 10^{-8} \text{ m}^2/\text{V·s} - 0.8 \times 10^{-8} \text{ m}^2/\text{V·s} = 2.3 \times 10^{-8} \text{ m}^2/\text{V·s}$

   The net flow is towards the negative electrode because the net mobility is positive.
   At pH 3.00, $\mu_{eo} = -1.2 \times 10^{-8} \text{ m}^2/\text{V·s} - 0.8 \times 10^{-8} \text{ m}^2/\text{V·s} = -2.0 \times 10^{-8} \text{ m}^2/\text{V·s}$
   The net flow is towards the positive electrode because the net mobility is negative.

104. The electrophoretic mobility (from Equation 21.13) of benzoate is

$$\mu_{ep,Bz} = \frac{\lambda_{Bz}}{zB_z F} = \frac{32.4 \text{ S cm}^2 \text{equiv}^{-1}}{-1 \times 96485 \text{ } coulombs \text{ equiv}^{-1}} = -3.36 \times 10^{-4} \text{ cm}^2 \text{V}^{-1} \text{s}^{-1}$$

The electrophoretic mobility of iodate is

$$\mu_{ep,103} = \frac{\lambda_{103}}{z_{103} F} = \frac{40.5 \text{ S cm}^2 \text{equiv}^{-1}}{-1 \times 96485 \text{ } coulombs \text{ equiv}^{-1}} = -4.20 \times 10^{-4} \text{ cm}^2 \text{V}^{-1} \text{s}^{-1}$$

The net mobility of benzoate is

$$\mu_{net,Bz} = \mu_{eo} + \mu_{ep,Bz} = (9.1 - 3.36) \times 10^{-4} \text{ cm}^2 \text{V}^{-1} \text{s}^{-1} = 5.7 \times 10^{-4} \text{ cm}^2 \text{V}^{-1} \text{s}^{-1}$$

The net mobility of iodate is

$$\mu_{net,103} = \mu_{eo} + \mu_{ep,IO3} = (9.1 - 4.20) \times 10^{-4} \text{ cm}^2 \text{V}^{-1} \text{s}^{-1} = 4.9 \times 10^{-4} \text{ cm}^2 \text{V}^{-1} \text{s}^{-1}$$

The migration time (Equation 21.25) of benzoate is

$$t_{Bz} = \frac{L^2}{\mu_{net,Bz} \times V} = \frac{(60 \text{ cm})^2}{5.7 \times 10^{-4} \text{ cm}^2 \text{V}^{-1} \text{s}^{-1} \times 2 \times 10^4 \text{ V}} = 316 \text{ s}$$

The migration time of iodate is

$$t_{IO3} = \frac{L^2}{\mu_{net,IO3} \times V} = \frac{(60 \text{ cm})^2}{4.9 \times 10^{-4} \text{ cm}^2 \text{V}^{-1} \text{s}^{-1} \times 2 \times 10^4 \text{ V}} = 367 \text{ s}$$

Based on Einstein Relation, Equation 21.12, the diffusion coefficient of benzoate

$$D_{Bz} = \frac{\mu_{ep,Bz} RT}{z_{Bz} F} = \frac{-3.36 \times 10^{-4} \text{ cm}^2 \text{V}^{-1} \text{s}^{-1} \times 8.314 \text{ J K}^{-1} \text{mol}^{-1} \times 298 \text{ K}}{-1 \times 96485 \text{ } coulombs \text{ equiv}^{-1}}$$

$$= 8.62 \times 10^{-6} \text{ cm}^2 \text{s}^{-1}$$

The diffusion coefficient of iodate

$$D_{IO3} = \frac{\mu_{ep,IO3} RT}{z_{IO3} F} = \frac{-4.20 \times 10^{-4} \text{ cm}^2 \text{V}^{-1} \text{s}^{-1} \times 8.314 \text{ J K}^{-1} \text{mol}^{-1} \times 298 \text{ K}}{-1 \times 96485 \text{ } coulombs \text{ equiv}^{-1}}$$

$$= 1.08 \times 10^{-5} \text{ cm}^2 \text{s}^{-1}$$

$$\mu_{net,Av} = \tfrac{1}{2} (5.7 + 4.9) \times 10^{-4} \text{ cm}^2 \text{ V}^{-1} \text{s}^{-1} = 5.3 \times 10^{-4} \text{ cm}^2 \text{V}^{-1} \text{s}^{-1}$$

$$D\mu ep = D\mu net = (5.7 - 4.9) \times 10^{-4} \text{ cm}^2 \text{V}^{-1} \text{s}^{-1} = 8 \times 10^{-5} \text{ cm}^2 \text{V}^{-1} \text{s}^{-1}$$

$$D_{Av} = \tfrac{1}{2} (8.6 + 1.08) \times 10^{-6} \text{ cm}^2 \text{s}^{-1} = 9.7 \times 10^{-6} \text{ cm}^2 \text{s}^{-1}$$

Applying Equation 21.30

$$R_s = 0.177 \, \Delta\mu_{ep} \sqrt{\frac{V}{\mu_{net,Av} D_{Av}}} = 0.177 \times 8 \times 10^{-5} \text{ cm}^2 \text{V}^{-1} \text{s}^{-1} \times \sqrt{\frac{2 \times 10^4 \text{ V}}{5.3 \times 10^{-4} \text{ cm}^2 \text{V}^{-1} \text{s}^{-1} \times 9.7 \times 10^{-6} \text{ cm}^2 \text{s}^{-1}}}$$

$$= 1.42 \times 10^{-5} \text{ cm}^2 \text{V}^{-1} \text{s}^{-1} \times 1.97 \times 10^6 \text{ cm}^{-2} \text{Vs} = 28$$

105. (a) Add some 18-crown-6 to the BGE: 18C6 complexes K^+ (but not NH_4^+) and increases its size and therefore will slow K^+ down, enabling the separation

(b) Increase the pH: Any change in EOF will affect both ions equally, this will not affect separation but increasing the pH will convert a greater portion of the total ammonium/ammonia into NH_3, which has no electrophoretic mobility. So in this case the movement of the NH_4^+/NH_3 will slow down and separation will improve.

(c) (Increase the voltage: Unless the voltage is already so high that heating is a problem, increasing the voltage will increase the efficiency and improve the separation.

(d) Use a longer capillary: Nothing, no improvement is expected. Unless a greater voltage can be applied to increase efficiency.

(e) Reverse the voltage polarity: The magnitude of EOF in this system is unknown. Since the pH is neutral EOF is likely positive. The cations are moving in the same direction as EOF. Reversing voltage will reverse both EOF and electrophoretic movement – these cations will never moves towards the detector!

(f) Use a smaller capillary: Nothing, there will be no gain unless the voltage was already too high and going toa smaller capillary will improve thermal management, separation efficiency and separation. If the voltage was previously not too high, then with a smaller capillary more voltage can be applied to improve separation. A smaller capillary will make detection limits poorer.

106. Using equation 21.34,

$$k = \frac{t_R - t_0}{t_0\left[1 - \left(\frac{t_R}{t_{MC}}\right)\right]}$$

t_R = Analyte time = 6.4 min
t_0 = Neutral marker time = 2.1 min
t_{MC} = Micelle time = 10.7 min

$$k = \frac{6.4\ min - 2.1\ min}{2.1\ min\left[1 - \left(\frac{6.4\ min}{10.7\ min}\right)\right]} = 5.1$$

107. Let us assume that the sample volume was V_s mL, the capillary cross section is x cm^2 and the sulfamate zone length is 1 cm. If we had injected 0.2 mM sulfamate and in the stacked zone the concentration is C_s mM, the law of mass conservation ($V_1C_1 = V_2C_2$) demands

$0.2\ V_s = C_s\ x$ (21.107.1)

or

$C_s = 0.2\ V_s/x$ (21.107.2)

Adapting equation 21.36

$$C_N = \frac{C_s \lambda_N (\lambda_S + \lambda_{Na})}{\lambda_s (\lambda_N + \lambda_{Na})} \quad\quad (21.107.3)$$

where subscripts s, N and Na, respectively, connote sulfamate, nitrate and sodium, and C_N represents the nitrate concentration in the stacked zone (in mM). If the sample nitrate concentration is $Sample_N$ mM, as we know the nitrate zone length is *0.5* cm, again law of mass conservation dictates

$$0.5 C_N \, x \, L = V_s \, Sample_N \quad\quad (21.107.4)$$

Rearranging
$$C_N = 2 \, V_s \, Sample_N / \, x \quad\quad (21.107.5)$$

Putting equations 21.107.2 and 21.107.3 in equation 21.107.5
$$2 \, V_s \, Sample_N = \frac{0.2 \, V_s x \lambda_N \, (\lambda_S + \lambda_{Na})}{x \lambda_s (\lambda_N + \lambda_{Na})} \quad\quad (21.107.6)$$

or

$$Sample_N = \frac{0.1 \lambda_N (\lambda_S + \lambda_{Na})}{\lambda_s (\lambda_N + \lambda_{Na})} \quad\quad (21.107.7)$$

From Table 21.2, λ_S, λ_N and λ_{Na} are, respectively, 48.6, 71.4 and 50.1 in the common units of S cm^2 /equiv., which cancels out in the ratio and is not further carried. Putting the λ values in equation 21.107.7, we get

$$Sample_N = \frac{0.1 * 71.4(48.6 + 50.1)}{48.6(71.4 + 50.1)} = 0.147 * (98.7/121.5) = 0.119 \text{ mM}$$

108. Plotting the square of the data gives better apparent efficiency and hence better apparent resolution of the peaks.

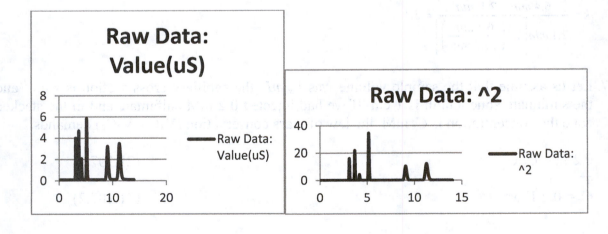

109. See the text website for the problem details and the spreadsheet. Use the regression analysis Excel program, as described in Section 16.7 of Chapter 16, and illustrated for an internal standard calibration in Section 20.5. Using the Excel spreadsheet Ch 20, Section 20.5 on Chapter 20 website, you can change the numbers for this problem, and for each sample. The unknown concentration is calculated from Variable X1 x area count ratio for the unknown. Variable X1 is 33.37993. The ratios for Samples 1, 2, and 3 are 1.055, 1.046, and 0.971, respectively. The calculated unknowns are 35.22, 34.92, and 32.41 ng/mL.

CHAPTER 22   MASS SPECTROMETRY

1. Inlet, Ionization Source, Mass Analyzer, Detector, Data Station, Vacuum System

2. Monoisotopic masses are calculated based on specified isotopes (usually the most abundant) for each element in a chemical compound. Mass spectrometers measure monoisotopic masses. Average masses are calculated from masses of elements weighted by the contributions from different naturally-occurring isotopes that have certain abundances.

3. In order to achieve high mass accuracy, sufficiently high resolution is needed to separate an ion from its nearest neighbor ion signals. Typically, the higher in resolution a mass analyzer can achieve, the greater the mass accuracy it can provide.

4. The molecular ion is given to the ion which represents the initial molecular form of a chemical compound. Generally, the molecule has been converted into its molecular ion by simply removing an electron or adding or removing a proton.

5. The rule that states if a chemical compound contains an even number (0, 2, 4,...) of nitrogens, then the molecular weight of the compound will be an even number. If it contains an odd number (1, 3, 5,...) of nitrogens, then the molecular weight of the compound will be an odd number.

6. Electron ionization (EI) and chemical ionization (CI).

7. Electrospray ionization (ESI) and atmospheric pressure chemical ionization (APCI).

8. A liquid containing analytes is flowed through a capillary that is held at a high potential. The electric field creates a mist of highly charged droplets that are sprayed into an atmospheric pressure chamber. As the droplets traverse the spray chamber, they reduce in size by solvent evaporation and droplet fission. Eventually, ions, primarily formed in solution, or ion-solvent clusters are released into the gas phase from the surface of the droplets, or remain after all solvent is removed. These ion species are then transferred into the high vacuum region of the mass spectrometer for mass analysis.

9. ESI can generate multiply-charged analytes, which means that high mass species can be given enough charges so that the ions can be manipulated by a standard mass analyzer with a given operational m/z ratio range.

10. APCI is a common ionization source for LC-MS, which induces ion formation as the vaporized solvent/analyte mixture is passed across a corona discharge needle to induce ion-molecule reactions at atmospheric pressure. Conventional CI is a common ionization source for GC-MS, where analytes enter an energetic electron ionization source, which has been saturated by a reagent gas that is ionized and can transfer charge to create analyte ions under vacuum.

11. The mean free path is the average distance a species can travel without colliding with another species. In mass spectrometry, it is important to protect formed ions from unwanted collisions, which might deactivate the ion. Therefore, the pressure in the mass spectrometer is reduced to increase the mean free path, typically to a point where the mean free path is 10- to 100-times the distance that the ion will have to travel in a given mass analyzer.

12. The Lorentz force ($F = qv\mathbf{B}$, where q is the charge of the ion, v is the velocity of the ion, and $\mathbf{B}$ is the strength of a magnetic field) describes the influence of an electric and magnetic field, assuming all are applied perpendicular to the flight path of the ion and each other, on an ion.

13. Single quadrupole instruments are mass filters that allow for fast scanning and a wide dynamic range for quantitative analysis. They provide only unit resolution and alone they cannot be used for tandem mass spectrometry instruments. The wider the range of m/z ratios desired to be monitored by a quadrupole instrument, the worse the sensitivity. However, they can be very useful for ion transfer (RF only mode) and they can be combined in a variety of ways with other instruments to create a hybrid instrument. Ion traps are inherently slower and they do not have a wide dynamic range. They also provide unit resolution (quadrupole ion traps), but they can be used to perform tandem mass spectrometry experiments on trapped ions. A wide range of m/z ratios can be trapped and analyzed without sacrificing sensitivity. Both single quadrupole and quadrupole ion trap instruments are quite affordable in comparison to other mass spectrometer configurations.

14. In FT-ICR, a homogeneous magnetic field is used to confine ions in a penning trap. The radius of curvature of a given ion in a magnetic field is dependent on its m/z ratio and the strength of the magnetic field. Thus, as higher magnetic field strengths are used, ions have larger m/z ratios can be confined in a given space.

15. A triple quadrupole instrument can be used to perform qualitative analysis by virtue of its capability for tandem mass spectrometry. An ion of interest can be isolated in the first quadrupole, fragmented in the second quadrupole, and the formed product ions scanned in the third quadrupole. This is called product ion scanning. Constant neutral loss and precursor ion scanning modes are also available techniques for gleaning qualitative information. For quantitative analysis, selected reaction monitoring (SRM) or multiple reaction monitoring (MRM) allow for highly sensitive and selective determinations. An ion of interest is selected in the first quadrupole, fragmented in the second quadrupole, and then a single fragment of the precursor ion is monitored. This reduces the possibility for interferences and allows all of the advantages of the quadrupole (fast duty cycle, wide dynamic range) to be utilized.

16. Caffeine has the elemental formula of $C_8H_{10}N_4O_2$. Adding up all of the masses for the most abundant isotopes of each element (carbon-12 = 12.00000 Da; hydrogen-1 = 1.00783; nitrogen-14 = 14.00307; oxygen-16 = 15.99491), a monoisotopic mass of 194.08040 Da is obtained. The protonated molecular ion ($[M+H]^+$) would thus have a monoisotopic mass of 194.08040 + 1.00783 = 195.08823 Da. The sodiated molecular ion ($[M+Na]^+$) would have a monoisotopic mass of 194.08040 + 22.98977 (sodium-23) = 217.07017 Da.

17. 2-Chlorobenzoic acid has the elemental formula of $C_7H_5ClO_2$. The monoisotopic mass for the $[M-H]^-$ ion would be 154.99052 Da. So, the predominant signal for the molecular ion would be at m/z 155 (100% relative intensity). The M+1 isotope signal (m/z 156) would be predominantly contributed from the fact that 1.1% of each carbon atom is carbon-13. Thus, the presence of 7 carbons gives the relative abundance of m/z 156 to be 7.7%. Since chlorine is present, a significant M+2 ion (25% of all chlorines, or one-third, are chlorine-37). Therefore, the presence of one chlorine in the molecule will give rise to a signal at m/z 157 with relative abundance of 33%. Other smaller contributions from all possible naturally-occurring isotopes can be considered to enumerate the abundances of other possible signals, but those above will be the three most abundant ion forms observed.

18. $R = m/\Delta m = 432.1124 / (432.1186 - 432.1124) = \underline{69,692}$. A resolution of approximately 70,000 would be needed to fully separate these two ion signals.

19. $R_{FWHM} = m/dm = 533.25 / (533.30 - 533.25) = \underline{10,665}$. Mass analyzers with moderate resolving power, such as reflectron time-of-flight or special operation modes of ion traps can achieve this resolution. High resolution instruments, such as FT-ICR or double-focusing magnetic sectors, could also achieve at least this level of resolution, but usually much greater values would be expected.

20. The m/z of the protein in the +34 charge state would be (34525 + 34) / 34 = 1016.4. The m/z of the protein in the +35 charge state would be (34525 + 35) / 35 = 987.4. $R = m/ \Delta m = 987.4 / (1016.4 - 987.4) = \underline{34}$. These monoisotopic ion signals could be easily resolved on a unit resolution instrument, however such a large molecule could have significantly abundant isotope signals or include various adducts that could broaden the peak for each charge state and complicate full resolution of charge state signals from one another.

21. ppm error in mass accuracy = $(m_{measured} - m_{true})/ m_{true}$ x 10^6 ppm = (1234.1198 – 1234.1223) / 1234.1223 x 10^6 = $\underline{-2.03}$ ppm. The error is negative because the measured mass was less than the true mass.

22. $KE = 0.75$ MeV $= 0.5mv^2$
0.75 x 10^6 eV x (1.602 x 10^{-19} J / 1 eV) = 1.2015 x 10^{-13} J [1 J = 1 kg m^2 s^{-2}]
m = 324.9 Da x (1.6605402 x 10^{-27} kg/Da) = 5.3951 x 10^{-25} kg

1.2015 x 10^{-13} J = 0.5 x 5.3951 x 10^{-25} kg x v^2
v = 667386 m s^{-1} = $\underline{6.67 \times 10^5}$ m s^{-1}

R = mv / qB = (5.3951 x 10^{-25} kg x 667386 m s^{-1}) / (1 x 1.602 x 10^{-19} C x 7 T) = 0.321 m = 32.1 cm

23. From $qV = 0.5mv^2$, we can obtain v = $(2zeV/m)^{0.5}$

For m/z = 1252.054
$v = (((2)(1)(1.602 \times 10^{-19} C)(20000V))/((1252.054 Da)(1.66 \times 10^{-27} kg/Da)))^{0.5}$
$v = 55525.9091$ m s^{-1}

$t = L/v = 1.750$ m $/ 55525.9091$ m s$^{-1} = 3.1516$ x 10^{-5} s $= 31.516$ µs

For m/z = 1253.138
$v = (((2)(1)(1.602$ x 10^{-19} C$)(20000$V$))/((1253.138$Da$)(1.66$x10^{-27} kg/Da$)))^{0.5}$
$v = 55501.8881$ m s^{-1}

$t = L/v = 1.750$ m $/ 55501.8881$ m s$^{-1} = 3.1530$ x 10^{-5} s $= 31.530$ µs

$\Delta t = 31.530 - 31.516 = 0.014$ µs $= \underline{14 \text{ ns}}$

CHAPTER 23   KINETIC METHODS OF ANALYSIS

1. The rate of a first-order reaction is proportional to the concentration of a single reactant. The rate of a second-order reaction is typically proportional to the product of the concentrations of the two reactants. (Strictly speaking, the order of a reaction is the sum of the exponents of the concentration terms in a rate equation.)

2. The half-life is the time it takes for a reaction to go to 50% completion. Theoretically, it takes an infinite number of half-lives to go to 100% completion, but practically speaking, a reaction can be considered essentially complete in about 10 half-lives.

3. A pseudo first-order reaction is a higher order reaction whose rate is made to depend on the concentration of a single reactant by making the concentration of the other reacting species high enough that they become essentially constant.

4. Set the starting concentrations of each equal. Plot $\log[A]$ against t (Equation 22.3). If the reaction is first order, a straight line will result, and if it is not, a curved line will result. Plot $([A]_0 - [A])/[A]$ vs. t (Equation 22.7). If the reaction is second order, a straight line will result.

5. An international unit (I.U.) is the amount of enzyme that will catalyze the transformation of one micromole of substrate per minute under defined conditions.

6. A competitive inhibitor competes with the substrate for an active site on the enzyme, and the inhibition varies with the concentration of the substrate. A noncompetitive inhibitor combines with the enzyme at a site other than the active site to form an inactive derivative of the enzyme, and the inhibition is independent of the concentration of the substrate.

7. They frequently combine with sulfhydryl groups in enzymes to inactivate the enzymes.

8. Substances which activate enzymes.

9. Double the substrate concentration and measure the rate of the reaction. If the inhibition is competitive, the percent inhibition will decrease, whereas if it is noncompetitive, it will remain the same.

10. If the inhibition is competitive, the slope of the plot will change, but the intercept will be the same. If the inhibition is noncompetitive, the intercept will change.

11. $t_{1/2} = (0.693/k)$
    $10.0 \text{ min} = (0.693/k)$
    $k = 0.0693 \text{ min}^{-1}$
    $\log[A] = \log[A]_0 - (kt/2.303)$

    Let $[A]_0 = 1$, then $[A]$ at 90% conversion $= 0.1$
    $\log 10^{-1} = \log 1 - (0.0693 \, t/2.303)$

t = 33.2 min

At 99% conversion, [A] = 0.01

$\log 10^{-2} = \log 1 - (0.0693t/2.303)$

t = 66.5 min

12. Let $[A]_0 = 1$, the [A] = 0.7 at 25.0 s

$\log 0.7 = \log 1 - (k \times 25.0)/2.303$

$k = 0.0143 \ s^{-1}$

$t_{1/2} = (0.693/0.0143) = 48.5 \ s$

13. $kt = ([A]_0 - [A])/([A]_0[A])$

$k \times 6.75 = [(0.100 - 0.850 \times 0.100)]/[0.100 \times (0.850 \times 0.100)]$

$k = 0.26 \ min^{-1}M^1$

$t_{1/2} = 1/(k[A]_0) = 1/(0.26 \ min^{-1}M^1 \times 0.100 \ M) = 38 \ min$

At 0.200 M:

$t_{1/2} = 1/(0.26 \times 0.200) = 19 \ min$

$0.26 \ t = [(0.200 - 0.850 \times 0.200)]/[0.200 \times (0.850 \times 0.200)]$

t = 3.4 min for 15.0% completion

14. At 25.0% conversion, [sucrose] = 0.750 x 0.500 M = 0.375 M

$\therefore \ \log 0.667 = \log 1 - (0.0320 \times t)/(2.303)$

t = 12.7 h

15. At 35%, the fraction of H_2O_2 remaining is 0.650.

$kt = ([A]_0 - [A]/[A]_0[A])$

$([A]_0 = [B]_0 = \frac{1}{2}[H_2O_2]_0)$

$k \times 8.60 = [(0.0500 - 0.650 \times 0.0500)]/[0.0500 \times (0.650 \times 0.0500)]$

$k = 1.25 \ min^{-1}M^1$

100 mL of O_2 at STP = (100 mL)/(22,400 mL/mol) = 0.00446 mol = 4.46 mmol O_2

Twice this many millimoles of H_2O_2 are used in forming the O_2 = 8.92 mmol

mmol H_2O_2 at start = 0.1000 M x 100.0 mL = 10.00 mmol

mmole H_2O_2 remaining = 10.00 – 8.92 = 1.08 mmol in 100 mL = 0.0108 M

$\therefore \ 1.25 \ t = [(0.0500 - 1/2 \times 0.0108)]/[0.0500 \times (1/2 \times 0.0108)]$

t = 132 h

16. μ mol O_2 consumed/20 min = (10.5 mL)/(0.0224 mL/μ mol) = 469 μ mol/20.0 min

= 23.4 μ mol/min

1 unit transforms 1 μ mol/min

$\therefore$ activity of enzyme preparation = 23.4 units/10.0 mg = 2.34 units/mg

% purity = (2.34/61.4) x 100% = 3.82%

17. See text website for spreadsheet solution.

Slope = 11.92 Δ A^{-1}min.mM

Intercept = $1/R_{max}$ = 9.925 Δ A^{-1}min

R_{max} = 0.1008 Δ A·min^{-1}

Uncertainy = $\pm$ 0.048 mM

K_m = slope x R_{max} = 1.20 $\pm$ 0.05 mM

CHAPTER 24   AUTOMATION IN MEASUREMENTS

1. An automatic instrument performs one or more steps of an analysis without human instruction. An automated instrument performs all steps of an analysis and uses the information to regulate a process without human intervention.

2. A continuous automated instrument continuously and constantly makes measurements on a process stream and feeds this information to controlling devices to continuously regulate the process. A discrete automated device makes measurements at discrete intervals to provide information to the regulator in discrete steps. The controlled variable is maintained at a fixed level between measurements.

3. A discrete sampling instrument analyzes each sample in a separate cuvet or chamber. A continuous flow sampling instruments analyzes the samples sequentially in a single tube. They are usually separated by air bubbles.

4. A feedback control loop utilizes measured information from a sensor about a variable to be controlled to compare against a set point in a controller. The controller feeds any difference to an operator which activates a regulating device to bring the variable back to the set point.

5. Flow injection analysis (FIA) is an unsegmented continuous flow technique in which a few microliters of sample are injected into a flowing reagent stream. Mixing occurs by diffusion and in about 15 s, product is detected as it flows through a micro-flow cell detector, resulting in a peak whose height is proportional to the sample concentration.

6. SIA is a single-line, single reversible pump system that uses a multiposition selection valve for aspirating reagent and sample plugs into a holding coil, then propels them to the detector. The adjacent zones merge and react on the way to the detector. It differs from FIA in that only a few microliters of reagent are used, and chemistries are readily changed by selecting a different reagent port, using the computer. Since it is not a continuous flow system, less waste is generated.

CHAPTER 25   CLINICAL CHEMISTRY

1. Serum, fibrinogen, and cells. Plasma contains serum and fibrinogen, and the cells include erythrocytes, leukocytes and platelets. When blood clots, the fibrinogen precipitates, removing the cells with it and leaving serum. When unclotted whole blood is centrifuged, the cells are separated from the plasma. Blood contains dozens of chemical constituents.

2. Hemolysis is the destruction of red cells, with the subsequent release of cellular constituents into the plasma or serum. The concentrations of a number of cellular constituents are higher than in the plasma and this would lead to erroneous results, if the plasma or serum is analyzed.

3. It acts as a glycolytic enzyme inhibitor to prevent the breakdown of glucose.

4. Because red cells are ruptured, releasing cellular constituents into the plasma.

5. Blood glucose and urea (BUN) analysis.

6. An antibody is a high molecular weight immunogloulin that specifically reacts with an antigen (foreign body).

7. The labels used in immunoassays include radiolabels, fluorescent labels, and enzyme labels.

8. Fluorescence immunoassays may be homogeneous or heterogeneous (requiring separation).

9. Enzyme immunoassays may employ competitive or noncompetitive binding.

CHAPTER 26   ENVIRONMENTAL SAMPLING AND ANALYSIS

1. Vacuum source, meter, collector

2. Impingers are used to sample aerosols in air.

3. Acidity, alkalinity, BOD, DO, conductivity, CO_2, Cl_2, F^-, NH_3, PO_4^{3-}, NO_3^-, SO_3^{2-}, metal ions, etc.

4. Protect from heat and light.

5. pH, dissolved gases, temperature.

6. For 1 mole, $V = (1 \text{ mol})(0.082 \text{ L atm/K mol})(293 \text{ K})/(1 \text{ atm}) = 24.0 \text{ L}$
    $(2.8 \times 10^{-6} \text{ L/L}_{air})/(24.0 \text{ L/mol}) = 1.1_7 \times 10^{-7} \text{ mol/L}_{air}$
    $(1.1_7 \times 10^{-7} \text{ mol/L}_{air})(28.0 \text{ g/mol}) = 3.3 \times 10^{-6} \text{ g/L}_{air}$
    7. $0.50 \ \mu \text{g/L} = 0.50 \text{ ppb} = 500 \text{ ppt}$
    $(97/128) \times 500 \text{ ppt} = 380 \text{ ppt toluene}$

CHAPTER G CENTURY OF THE GENE—GENOMICS AND PROTEOMICS: DNA SEQUENCING AND PROTEIN PROFILING

1. Chromosomes are made up of thousands of genes. There are 23 chromosome pairs in the nucleus of each cell (except sperm and egg cells).

2. DNA is a helical polymer consisting of nucleotides with four different nucleic acid bases, adenine (A), cytosine (C), guanine (G), and thymine (T). The bases A and T, and G and C, pair by hydrogen bonding to form the "steps" of the helical DNA.

3. The polymerase chain reaction (PCR) is a technique for replicating trace quantities of DNA. The DNA is heated and cooled through several cycles in the presence of the four nucleotides, a primer, and a DNA polymerase. The heating causes the DNA strands to separate (denature). The polymerase catalyzes extension of the primer to produce a complement stand of DNA template, thereby doubling the number of DNA molecules in each cycle.

4. Plasmids and bacterial artificial chromosomes (BACs) are vectors in which DNA fragments are inserted. These are put in bacteria where they are replicated. Plasmids insert relatively small DNA fragments (2-20 kb), while BACs insert large fragments (100-400 kb).

5. A genomic library is a set of BAC replicate DNA fragments, from a partially digested genome. These are provided to DNA sequencing laboratories, which further digest them, replicate the smaller fragments, and then determine the base sequence in the overlapping fragments.

6. BAC clones of up to 300 kb are further fragmented using nucleases, to give strands of 100-300 bases. These are replicated by insertion into plasmids. The plasmid clones are subjected to PCR replication in the presence of dideoxynucleotides, which results in complementary strands of every length, terminated by one of four dideoxynucleotides. These are separated in order of length using electrophoresis, and identified by the color of the end nucleotide fluorescence. Overlapping fragments are aligned to identify the entire DNA sequence.

7. A SNP is a single nucleotide polymorphism. These represent about 0.1% of the genes that result in the features that make us all different, and contribute to disease and dysfunction. These genes differ by a single nucleotide.

8. DNA chips are known single strand DNA fragments that are tethered to a plate, in the form of a microarray. Unknown DNA fragment samples are identified from the fluorescent spots created when the unknown binds to its complement tether.

9. Expression profiling is the study of gene expression by comparing mRNA sequences between known and unknown samples, for example, comparing a patient's expression with that of a known disease. DNA (RNA) chips are used to identify the mRNA sequences.

10. Genomics is the study of the DNA expression (encoding) process, to form proteins in the cell. Proteomics is the study of the complements of protein structure and function in a cell.

11. Proteins consist of a polymeric sequence of amino acids, derived from twenty different amino acids.

12. The gene DNA is transcribed into an m-RNA in the cell nucleus, which migrates into the cytoplasm where it serves as the template for protein production; 3-base codons encode for a particular amino acid.

13. 2-D PAGE is 2-dimensional polyacrylamide gel electrophoresis. It separates proteins, first by isoelectric focusing based on charge, and then by gel electrophoresis based on molecular size.

14. MALDI-TOF is matrix assisted laser desorption ionization-time-of-flight mass spectrometry. It is a soft ionization technique for large molecules that produces single peaks from singly charged molecules. It is used to identify the masses of peptides in protein digestion mixtures.

15. Theoretical protein amino acid fingerprints are constructed from a combination of a knowledge of the gene codons and a large protein database. The experimental fingerprint is compared with the theoretical ones to identify the protein from fingerprints of as few as 5 or 6 peptides.

16. The first sequence is:
    GATCCA**ATTGCAT**

    The second is:
    GATCCA**CATTCCGTA**

    They overlap as follows:
    GATCCA**ATTGCAT**
    GATCCA**CATTCCGTA**

    The sequence is:
    ATTGCATTCCGTA